Applied Aspects of Optical Communication and LIDAR

OTHER TELECOMMUNICATIONS BOOKS FROM AUERBACH

Broadband Mobile Multimedia:
Techniques and Applications
Yan Zhang, Shiwen Mao, Laurence T. Yang,
and Thomas M Chen
ISBN: 978-1-4200-5184-1

Carrier Ethernet: Providing the Need for Speed
Gilbert Held
ISBN: 978-1-4200-6039-3

Cognitive Radio Networks
Yang Xiao and Fei Hu
ISBN: 978-1-4200-6420-9

Contemporary Coding Techniques and
Applications for MobileCommunications
Onur Osman and Osman Nuri Ucan
ISBN: 978-1-4200-5461-3

Converging NGN Wireline and Mobile 3G
Networks with IMS: Converging NGN and
3G Mobile
Rebecca Copeland
ISBN: 978-0-8493-9250-4

Cooperative Wireless Communications
Yan Zhang, Hsiao-Hwa Chen, and Mohsen Guizani
ISBN: 978-1-4200-6469-8

Data Scheduling and Transmission Strategies
in Asymmetric Telecommunication
Environments
Abhishek Roy and Navrati Saxena
ISBN: 978-1-4200-4655-7

Encyclopedia of Wireless and Mobile
Communications
Borko Furht
ISBN: 978-1-4200-4326-6

IMS: A New Model for Blending Applications
Mark Wuthnow, Jerry Shih, and Matthew Stafford
ISBN: 978-1-4200-9285-1

The Internet of Things: From RFID to the
Next-Generation Pervasive Networked
Systems
Lu Yan, Yan Zhang, Laurence T. Yang,
and Huansheng Ning
ISBN: 978-1-4200-5281-7

Introduction to Communications
Technologies: A Guide for Non-Engineers,
Second Edition
Stephan Jones, Ron Kovac, and Frank M. Groom
ISBN: 978-1-4200-4684-7

Long Term Evolution: 3GPP LTE Radio
and Cellular Technology
Borko Furht and Syed A. Ahson
ISBN: 978-1-4200-7210-5

MEMS and Nanotechnology-Based Sensors
and Devices for Communications,
Medical and Aerospace Applications
A. R. Jha
ISBN: 978-0-8493-8069-3

Millimeter Wave Technology in Wireless PAN,
LAN, and MAN
Shao-Qiu Xiao and Ming-Tuo Zhou
ISBN: 978-0-8493-8227-7

Mobile Telemedicine: A Computing and
Networking Perspective
Yang Xiao and Hui Chen
ISBN: 978-1-4200-6046-1

Optical Wireless Communications:
IR for Wireless Connectivity
Roberto Ramirez-Iniguez, Sevia M. Idrus,
and Ziran Sun
ISBN: 978-0-8493-7209-4

Satellite Systems Engineering in an
IPv6 Environment
Daniel Minoli
ISBN: 978-1-4200-7868-8

Security in RFID and Sensor Networks
Yan Zhang and Paris Kitsos
ISBN: 978-1-4200-6839-9

Security of Mobile Communications
Noureddine Boudriga
ISBN: 978-0-8493-7941-3

Unlicensed Mobile Access Technology:
Protocols, Architectures, Security,
Standards and Applications
Yan Zhang, Laurence T. Yang, and Jianhua Ma
ISBN: 978-1-4200-5537-5

Value-Added Services for Next Generation
Networks
Thierry Van de Velde
ISBN: 978-0-8493-7318-3

Vehicular Networks: Techniques, Standards,
and Applications
Hassnaa Moustafa and Yan Zhang
ISBN: 978-1-4200-8571-6

WiMAX Network Planning and Optimization
Yan Zhang
ISBN: 978-1-4200-6662-3

Wireless Quality of Service:
Techniques, Standards, and Applications
Maode Ma and Mieso K. Denko
ISBN: 978-1-4200-5130-8

AUERBACH PUBLICATIONS
www.auerbach-publications.com
To Order Call: 1-800-272-7737 • Fax: 1-800-374-3401
E-mail: orders@crcpress.com

Applied Aspects of Optical Communication and LIDAR

Nathan Blaunstein
Shlomi Arnon
Arkadi Zilberman
Natan Kopeika

CRC Press
Taylor & Francis Group
Boca Raton   London   New York

CRC Press is an imprint of the
Taylor & Francis Group, an **informa** business
AN AUERBACH BOOK

Auerbach Publications
Taylor & Francis Group
6000 Broken Sound Parkway NW, Suite 300
Boca Raton, FL 33487-2742

© 2010 by Taylor and Francis Group, LLC
Auerbach Publications is an imprint of Taylor & Francis Group, an Informa business

No claim to original U.S. Government works

Printed in the United States of America on acid-free paper
10 9 8 7 6 5 4 3 2 1

International Standard Book Number: 978-1-4200-9040-6 (Hardback)

Library of Congress Cataloging-in-Publication Data

Applied aspects of optical communication and LIDAR / authors, Nathan Blaunstein ... [et al.].
    p. cm.
    "A CRC title."
    Includes bibliographical references and index.
    ISBN 978-1-4200-9040-6 (hardcover : alk. paper)
    1. Free space optical interconnects. 2. Optical communications. 3. Optical radar. 4. Light--Transmission. 5. Meteorological optics. I. Blaunstein, Nathan. II. Title.

TK5103.592.F73A67 2010
621.382'7--dc22
                                                                    2009038082

Visit the Taylor & Francis Web site at
http://www.taylorandfrancis.com

and the Auerbach Web site at
http://www.auerbach-publications.com

Contents

Preface

This book deals with different aspects of optical wave propagation phenomena in the atmosphere, mostly close to the terrain (up to 10–20 km), and describes the main important characteristics of optical propagation in wireless communication links and their applications to LIDAR (light detection and ranging) systems. Nowadays, we observe a conversion of existing radio networks (labeled as the second generation, or 2G) and third generation (3G) to combined optical-radio wireless networks with high rates of information sent via such channels and a small bit error rate (BER). To design these networks successfully, it is very important to predict the *propagation characteristics* of optical atmospheric wireless communication links in order to provide high quality of service (QoS) for any individual subscriber located in a given area of service.

These functions can be achieved by using strict prediction of the key parameters of optical communication links at one or both ends of a communication link. Accurate information about the physical propagation processes that occur in each atmospheric-specific scenario increases the performance of optical communication links and the efficiency of service (called grade of service, GoS) for each subscriber.

This book involves a synthesis of different backgrounds in order to present a broad and unified approach to propagation in different scenarios occurring in the atmosphere such as turbulence, aerosols, rain, snow, clouds, and so on.

This book is intended for any graduate, postgraduate student, scientist, practicing engineer, or designer who is concerned with the operation and service of atmospheric optical links, mostly for laser and lidar applications in atmospheric optical communication links. It examines different situations in over-the-terrain atmospheric communication channels, including the effects of atmospheric turbulence and different kinds of hydrometeors (rain, clouds, snow, etc.) on atmospheric links. For each specific situation the main task of this book is to explain the effects of all kinds of natural phenomena and the corresponding features (e.g., turbulences and hydrometeors) on optical ray propagation that, finally, influence the transmission of optical signals through perturbed atmospheric communication channels, in both line-of-sight (LOS) and obstructive non-line-of-sight (NLOS) propagation conditions. The book also emphasizes how to use lidars to investigate atmospheric phenomena and predict primary parameters of atmospheric optical channels, and to suggest what kinds of optical devices can be utilized to minimize the deleterious effects of natural atmospheric phenomena. It also serves as a bridge between parameters of optical communication links and signal information data streams propagating through perturbed atmospheric channels.

The book introduces the reader to some relevant topics of optical wave propagation in the turbulent atmosphere and their relations to laser and lidar applications. Multipath phenomena, path loss, large-scale or slow fading, and short-scale or fast fading are thoroughly described. The phenomena treated include free-space optical propagation above the terrain, propagation in the inhomogeneous and stratified atmosphere, and reflection and diffraction by various obstructions (aerosols, hydrometeors, and turbulences) regularly or randomly distributed along the link of communication. Finally, the authors try to show how to create a unified approach for predicting key characteristics for different optical wireless communication channels and the main parameters of the data stream (capacity, spectral efficiency, and BER). That means a full prediction of key propagation characteristics and signal information data parameters for various atmospheric optical communication links, without which the proposed prediction frameworks cannot be used successfully.

The organization of the book is as follows. The main features and phenomena occurring in the inhomogeneous atmosphere as well as the key parameters and characteristics of the atmospheric propagation links are briefly described in Chapter 1. Chapter 2 introduces the reader to the physics of optical wave propagation in inhomogeneous atmospheric media, based on classical and modern theoretical principles, some of which were investigated by the authors of this book, for applications in optical propagation above the terrain at altitudes not higher than 20 km, emphasizing mainly the effects of turbulent atmosphere on optical ray propagation. In Chapter 3, the authors present the principal aspects of lidars for studying the main features of the turbulent stratified atmosphere and how, based on obtained experimental data and the corresponding empirical and semi-empirical models, to predict the key parameters of optical atmospheric channels. All aspects of optical communication networks, based on knowledge of propagation phenomena mentioned in previous chapters, are covered in Chapter 4. Here, a general mathematical tool of how to predict key parameters of optical signals within the perturbed atmospheric channel are presented based on general statistical theory and general probability functions describing different scenarios occurring in atmospheric communication channels. Finally, in Chapter 5, a general statistical approach, presented in Chapter 4, is used to suggest prediction algorithms for how to obtain signal data parameters such as capacity, spectral efficiency, and BER for different situations occurring in the inhomogeneous turbulent atmosphere, with examples for several specific atmospheric communication links.

Acknowledgments

The authors are grateful for the contributions of Dr. Ephim Golbraikh in the theoretical analysis of various turbulence phenomena such as non-Kolmogorov turbulence and its effects on electromagnetic wave propagation mentioned in Chapters 1 and 2.

The authors also acknowledge the experimental work and empirical modeling contributed by former student Sergey Bendersky, regarding the effects of various turbulence phenomena on laser beam propagation, described in Chapter 2, as well as former students Alexander Tiker and Dr. Nathalie Yarkoni for their contributions in modeling of signal data parameters in optical atmospheric channels, presented in Chapter 5.

In addition, Professor Arnon would like to thank his former student Dr. Debbie Kedar for valuable discussion and proofreading during the writing of Chapter 4.

Some of the research described here was supported with a grant from the Israel Science Foundation (grant no. 730/06).

We are greatly indebted to our families for providing the wonderful environment in which this work was performed, and to the Taylor & Francis staff for their superb editorial work.

About the Authors

Professor Shlomi Arnon is a faculty member in the Department of Electrical and Computer Engineering at Ben-Gurion University (BGU), Israel. There, in 2000, he established the Satellite and Wireless Communication Laboratory, which has been under his directorship since then. During 1998–1999, Professor Arnon was a postdoctoral associate (Fulbright fellow) at LIDS, Massachusetts Institute of Technology (MIT), Cambridge, USA. His research has produced more than 50 journal papers in the areas of satellite, optical, and wireless communication. During part of the summer of 2007, he worked at TU/e and PHILIPS LAB, Eindhoven, the Netherlands, on a novel concept of a dual communication and illumination system. He was visiting professor during the summer of 2008 at TU Delft, the Netherlands.

Professor Arnon is a frequent invited speaker and program committee member at major IEEE and SPIE conferences in the United States and Europe. He was an associate editor for the Optical Society of America's *Journal of Optical Networks*, for a special issue on optical wireless communication that appeared in 2006, and he is now on the editorial board for the *IEEE Journal on Selected Areas in Communications* for a special issue on optical wireless communication.

Professor Arnon continuously takes part in many national and international projects in the areas of satellite communication, remote sensing, and cellular and mobile wireless communication. He consults regularly with start-up and well-established companies in the areas of optical, wireless, and satellite communication. In addition to research, Professor Arnon and his students work on many challenging engineering projects with special emphasis on the humanitarian dimension. For instance, long-standing projects have dealt with developing a system to detect human survival after earthquakes and a respiration monitoring system to prevent infant cardiac arrest and apnea or detect falls in the case of epilepsy sufferers and elderly people.

Nathan Blaunstein received BS and MS degrees in radiophysics and electronics from Tomsk University, Tomsk, former Soviet Union, in 1972 and 1975, respectively, and PhD and DS and professor degrees in radiophysics and electronics from the Institute of Geomagnetism, Ionosphere, and Radiowave Propagation (IZMIR), Academy of Science USSR, Moscow, Russia, in 1985 and 1991, respectively. From 1979 to 1984, he was an engineer and a lecturer, and then, from 1984 to 1992, a senior scientist, an associate professor, and a professor at Moldavian University, Beltsy, Moldova, former USSR. From 1993 he was a senior scientist of the Department of Electrical and Computer Engineering and a visiting professor in the Wireless Cellular Communication Program at the Ben-Gurion University of the Negev, Beer-Sheva, Israel. Since April 2001, he has been a professor in the Department of Communication Systems Engineering.

Professor Blaunstein has published 6 books, 2 special chapters in handbooks on applied engineering and applied electrodynamics, 6 manuals, and more than 160 articles on radio and optical physics, communication, and geophysics.

His research interests include problems of radio and optical wave propagation, diffraction, reflection, and scattering in various media (sub-soil medium, terrestrial environments, troposphere, and ionosphere) for purposes of optical communication and radio location, aircraft, mobile-satellite, and terrestrial wireless communications and networking.

N. S. Kopeika was born in Baltimore in 1944. Raised in Philadelphia, he received BS, MS, and PhD degrees in electrical engineering from the University of Pennsylvania in 1966, 1968, and 1972, respectively.

He and his family moved to Israel, and he began his career at Ben Gurion University of the Negev in 1973. He chaired the Department of Electrical and Computer Engineering for two terms (1989–1993) and was named Reuven and Francis Feinberg Professor of Electrooptics in 1994. He was the first head of the Department of Electrooptical Engineering (2000–2005), which grants graduate degrees in electrooptical engineering.

He has published more than 170 papers in international reviewed journals and well over 100 papers at various conferences. Recent research involves development of a novel inexpensive focal plane array camera for terahertz imaging. He is a fellow of SPIE — The International Society for Optical Engineering. Other areas of research include interactions of electromagnetic waves with plasmas, the optogalvanic effect, environmental effects on optoelectronic devices, imaging system theory, propagation of light through the atmosphere, imaging through the atmosphere, image processing and restoration from blur, imaging in the presence of motion and vibration, lidar, target acquisition, and image quality in general. He is the author of the textbook *A System Engineering Approach to Imaging*, published by SPIE Press (first printing 1998, second printing 2000), and is a topic editor for Marcel Dekker for "atmospheric optics" in their *Encyclopedia of Optical Engineering*.

Arkadi Zilberman received BS and MS degrees in physics from Tomsk State University, Tomsk, former Soviet Union, in 1994, and an MS degree in electrical engineering from Ben-Gurion University of the Negev, Israel, in 2001. He received a PhD degree from Ben-Gurion University of the Negev, Israel, in 2006.

From 1994 until 1996 he was a researcher at the Institute of Atmospheric Optics, Tomsk, former Soviet Union. Currently, he is a researcher in the Department of Electrical and Computer Engineering at the Ben-Gurion University of the Negev, Beer-Sheva, Israel.

Dr. Zilberman has published more than 30 papers in various journals and conference proceedings concerning optical and atmospheric physics, lidar, and optical imaging. His research interests are optical wave propagation, atmospheric communication, lidar, and imaging.

1

ATMOSPHERE

Structure and Processes

The atmosphere is a gaseous envelope that surrounds the Earth from the ground surface up to several hundred kilometers. The atmosphere comprises different kinds of gaseous, liquid, and crystal structures, including the effects of gas molecules (atoms), aerosol, cloud, fog, rain, hail, dew, rime, glaze, and snow [1–16]. Except for the first two, they are usually called in the literature *hydrometeors* [9–15]. Furthermore, due to irregular and sporadic air streams and motions — that is, irregular wind motions — the chaotic structures defined as *atmospheric turbulence* are also present in the atmosphere [17–21].

Mostly on the basis of temperature variations, the Earth's atmosphere is divided into four primary layers [20]: *troposphere* (up to ~11 km altitude) with tropopause region (isothermal layer above troposphere up to ~20 km); *stratosphere* with stratopause (from 20 km up to ~50 km altitude); *mesosphere* with mesopause (up to ~90 km); and *thermosphere* (up to ~600 km). Most of the ionosphere is included in the thermosphere.

The physical properties of the atmosphere are characterized by main parameters such as *temperature*, T (in Kelvin); *pressure*, P (in millibars, pascals, or mm Hg), and *density*, ρ (in kg m^{-3}). All these parameters significantly change with altitude as well as seasonal and latitudinal variability, and strongly depend on weather [22]. The following section provides a brief introduction to the main kinds of gaseous, liquid, and crystal structures in the atmosphere.

1.1 Vertical Profiles of Temperature, Pressure, and Number Density

Over 98% of the atmosphere by volume is comprised of the elements nitrogen and oxygen. The number density of nitrogen molecules, $\rho_N(h)$, at height h can be found in the *U.S. Standard Atmosphere* [22].

The temperature $T(h)$, in degrees Kelvin, and pressure $P(h)$, in pascals, as a function of the altitude h, in meters, for the first 11 km of the atmosphere can be determined from the following expressions [23]:

$$T(h) = 288.15 - 65.45 \cdot 10^{-4} h \tag{1.1}$$

$$P(h) = 1.013 \times 10^5 \cdot \left[\frac{288.15}{T(h)} \right]^{5.22} \tag{1.2}$$

The temperature and pressure from 11 to 20 km in the atmosphere can be determined from:

$$T(h) = 216.65 \tag{1.3}$$

$$P(h) = 2.269 \times 10^4 \cdot \exp\left[-\frac{0.034164(h - 11000)}{216.65} \right] \tag{1.4}$$

The temperature and pressure from 20 to 32 km in the atmosphere can be determined from:

$$T(h) = 216.65 + 10^{-3} \cdot (h - 20000) \tag{1.5}$$

$$P(h) = 5528.0 \cdot \left[\frac{216.65}{T(h)} \right]^{34.164} \tag{1.6}$$

The temperature and pressure from 32 to 47 km in the atmosphere can be determined from:

$$T(h) = 228.65 + 0.0028 \cdot (h - 32000) \tag{1.7}$$

$$P(h) = 888.8 \cdot \left[\frac{228.65}{T(h)} \right]^{12.201} \tag{1.8}$$

The number density of molecules can be found from:

$$\rho(h) = \left(\frac{28.964 \, \text{kg/kmol}}{8314 \, \text{J/kmol} - K} \right) \cdot \frac{P(h)}{T(h)} = 0.003484 \cdot \frac{P(h)}{T(h)} \, \text{kg/m}^3 \tag{1.9}$$

1.2 Aerosols

Atmospheric aerosols comprise a dispersed system of small solid and liquid particles suspended in air. Remaining suspended for varying periods of time, they are transported by vertical and horizontal wind currents, frequently to great distances. Aerosols are formed by two main processes: a primary source, which includes dispersion of particulates from the Earth's surface (like soil and deserts, oceans, volcanoes, biomass burning, industrial injection), and a secondary source resulting from atmospheric chemical reactions, condensation, or coagulation processes [23–34]. Aerosol concentrations and properties depend on the intensity of the sources, on the atmospheric processes that affect them, and on the particle transport from one region to another. The size distribution of the atmospheric aerosol is one of its core physical parameters. It determines the various properties such as mass and number density, or optical scattering, as a function of particle radius. For the atmospheric aerosols, this size range covers more than five orders of magnitude, from about ten nanometers to several hundred micrometers. This particle size range is very effective for scattering of radiation at ultraviolet (UV), visible, and infrared (IR) wavelengths. The aerosol size distribution varies from place to place, with altitude, and with time.

In a first attempt to sort into geographically distinct atmospheric aerosols, Junge classified aerosols, depending on their location in space and sources, into background, maritime, remote continental, and rural [24]. This classification later was expanded and quantified [25].

Some studies have been devoted to specific aerosol types: desert aerosols [26, 27], urban aerosols [28], aerosols resulting from biomass burning in tropical regions [29], and stratospheric aerosols [30]. Of course all classification models reflect only certain average values. Individual distributions vary depending on local weather and wind, vertical mixing, horizontal transport, gas-to-particle conversion, season, and so on.

Of all the classification parameters for atmospheric aerosols, the one most commonly used is size. General classification suggests three modes of aerosols [31]: (1) a *nuclei* mode, which is generated by spontaneous nucleation of the gaseous material for particles less than 0.1 μm in diameter; (2) the *accumulation* mode, for particles between 0.1 and 1 μm diameter, mainly resulting from coagulation and in cloud processes; and (3) the *coarse* mode, for particles larger than 1.0 μm in diameter originating from the Earth's surface (land and ocean). This classification is quite similar to Junge's designation [24], referred to as Aitken, large, and giant particles. The particles vary not only in chemical composition and size but also in shape (spheres, ellipsoids, rods, etc.).

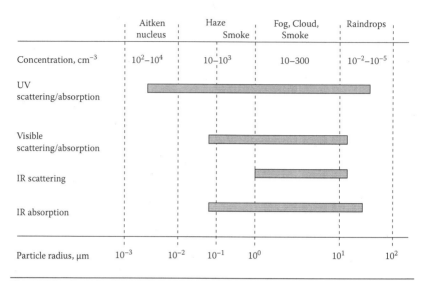

Figure 1.1 Aerosol types and their interaction with radiation.

From the optical standpoint, to underline the scattering properties of the atmosphere the term *haze aerosol* was introduced [32, 33]. Hazes are polydisperse aerosols in which the size range of particles extends from about 0.01 to 10 µm. Haze is a condition wherein the scattering property of the atmosphere is greater than that attributable to the gas molecules but is less than that of fog. Haze can include all types of aerosols. Cosmic dust, volcanic ash, foliage exudations, combustion products, bits of sea salt — all these are found to varying degrees in haze.

The size range for different types of aerosols and their interaction with radiation are summarized in Figure 1.1.

1.2.1 Aerosol Loading

Because the aerosol in the atmosphere exhibits considerable variation in location, height, time, and constitution, different concepts exist for describing the aerosol loading in the atmosphere.

Models for the vertical variability of atmospheric aerosols are generally broken into a number of distinct layers. In each of these layers a dominant physical mechanism determines the type, number density,

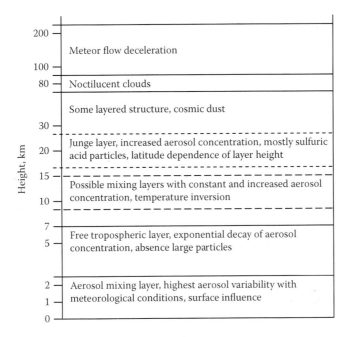

Figure 1.2 Vertical variability of atmospheric aerosols.

and size distribution of particles. Generally accepted layer models consist of the following [34, 35]: a boundary layer that includes aerosol mixing goes from 0 to 2–2.5 km elevation; a free tropospheric region runs from 2.5 to 7–8 km; a stratospheric layer from 8 to 30 km; and layers above 30 km are composed mainly of particles that are extraterrestrial in origin, such as meteoric dust [21]. Figure 1.2 is an example of such a description.

The average thickness of the aerosol-mixing region is approximately 2–2.5 km. Within this region, one would expect the aerosol concentration to be influenced strongly by conditions at ground level. Consequently, aerosols in this region display the highest variability with meteorological conditions, climate, etc. [35–42]. Variability of aerosol number concentration at different locations can be seen from measurements. For example, the typical aerosol number concentrations in an urban area of the Negev desert (city of Beer-Sheva, Israel) ranged from 1.5×10^3 to 5×10^3 cm^{-3} for normal conditions and exceeded 3×10^4 cm^{-3} for disturbed conditions (particle radii from 0.16 to 10 μm).

In the free tropospheric layer that extends from 2–2.5 to 7–8 km, an exponential decay of aerosol number density is observed. One often sees total number densities vary as follows [12]:

$$N(z) = N(0)\exp\left(-\frac{z}{z_s}\right), \tag{1.10}$$

where the scale height, z_s, ranges from 1 to 1.4 km. It can be taken in most cases to equal 1.21. Another proposed form of $N(z)$ takes into account an inversion layer (z_1) and mixing turbulent layer (z_2, boundary layer) and is given by [36]

$$N(z) = \begin{cases} N(z_0) - \left[\dfrac{N(z_0) - N(z_1)}{z_1}\right] \cdot z, & z_0 \leq z \leq z_1 \\[2ex] N(z_1) = const, & z_1 < z \leq z_2 \\[2ex] N(z_2)\exp\left[\dfrac{z_2 - z}{8}\right], & z > z_2 \end{cases} \tag{1.11}$$

where the heights z_1 and z_2 for a semi-arid zone are $z_1 = 0.3$ km and $z_2 = 4$ km for summer, and $z_1 = 0.4$ km and $z_2 = 1.5$ km for winter.

The few data available show that in the atmospheric boundary layer (0–2 km) and in the lower stratosphere (9–14 km altitude) there may exist different layers (known as mixing layers) with constant and increased aerosol concentration [37, 38]. These layers can be caused by temperature inversions at ground level and by tropopause effects where the temperature gradient changes sign. Moreover, in the stratospheric region there is some latitude dependence of aerosol layer height (Junge layer). The concentration maximum of stratospheric aerosols near the equator is located at 22–26 km elevation altitude, but at about 17–18 km height in the polar region.

The U.S. Air Force Geophysical Laboratory (AFGL) has developed a model that describes transmittance losses due to particulate absorption and scattering: MODTRAN [39, 40]. In MODTRAN, the variation of the aerosol optical properties with altitude is modeled by dividing the atmosphere into four height regions each having a different type of aerosol. These regions are the boundary and mixing layer (0–2 km), the upper troposphere (2–10 km), the lower stratosphere (10–30 km), and the upper atmosphere (30–100 km). The models for the troposphere (rural, urban, maritime, and tropospheric) are given as a function of the relative humidity [34]. A wind-dependent desert aerosol model and background stratospheric aerosol model have been introduced [27]. d'Almeida et al. [16] have compiled various studies and developed a global climatology of aerosol optical properties for Arctic, Antarctic, desert, continental, urban, and maritime regions. Using these aerosol distributions, they have modeled aerosol optical properties on a global scale.

1.2.2 Aerosol Size Distribution and Spectral Extinction

Atmospheric aerosols are polydisperse. Although their sizes can vary in the 0.01–100 μm range of diameters, the number of small particles is limited by the coagulation process and the number of large particles by gravitational sedimentation. Between these limits, the number of particles varies with size.

The manner in which the particle population is spread over the range of sizes is defined by the size distribution function. The size distribution of the atmospheric aerosol is one of its core physical parameters. It determines how the various properties, such as mass and number density or optical scattering, are distributed as a function of particle radius. Particle size distributions are necessary inputs for models used to predict the attenuation and scatter of radiation between the transmitter and receiver in different applications (optical communication, satellite image restoration, weapons-based electro-optical systems, etc.).

The number, $n(r)$, of particles per unit interval of radius and per unit volume is given by

$$n(r) = \frac{dN(r)}{dr}. \tag{1.12}$$

The differential quantity $dN(r)$ expresses the number of particles having a radius between r and $r + dr$, per unit volume, according to the distribution function $n(r)$.

Because of the many orders of magnitude present in atmospheric aerosol concentrations and radii, a logarithmic size distribution function is often used:

$$n(r) = \frac{dN(r)}{d\log(r)}. \tag{1.13}$$

The much-used distribution function is the power law first presented by Junge [24, 41]. Junge's model is

$$n(r) = \frac{dN(r)}{d\log r} = Cr^{-\upsilon}, \quad r \geq r_{\min} \tag{1.14}$$

or, in a non-logarithmic form,

$$n(r) = \frac{dN(r)}{dr} = 0.434Cr^{-\upsilon}, \tag{1.15}$$

where C is the normalizing constant to adjust the total number of particles per unit volume and υ is the shaping parameter. Most measured size distributions can best be fit by values of υ in the range $3 \leq \upsilon \leq 5$, for hazy and clear atmospheric conditions, and for aerosols whose radii lie between 0.1 and 10 µm [42]. According to the power law size distribution, the number of particles decreases monotonically with an increase in radius. In practice, there is an accumulation in the small particle range. Actual particle size distributions may differ considerably from a strict power law form.

McClatchey et al. [42] provided a modified power law distribution, according to which the number of particles within the radius interval dr is given by

$$n(r) = C \qquad \text{for } 0.02\ \mu\text{m} < r < 0.1\ \mu\text{m},$$

$$n(r) = Cr^{-\upsilon} \quad \text{for } 0.1\ \mu\text{m} < r < 10\ \mu\text{m}, \tag{1.16}$$

$$n(r) = 0 \qquad \text{for all other values of } r$$

The power law distribution does not show modulations in the shape of the distribution. To allow for the drop-off of particle number density at small radii, the modified gamma distribution function was introduced. It has the form [43]:

$$n(r) = \frac{dN(r)}{dr} = ar^{\alpha}\exp(-br^{\beta}), \tag{1.17}$$

where a is the total number density, and α, β, and b are the shaping parameters. The total particle concentration, given by the integral over all particle radii, for this distribution is [40]

$$N = a\beta^{-1}b^{-(\alpha+1)/\beta}\Gamma\left(\frac{\alpha+1}{\beta}\right),$$ (1.18)

The mode radius for this distribution is given by

$$r_m^{\beta} = \frac{\alpha}{b\beta}.$$ (1.19)

The value of the distribution at the mode radius is

$$n(r_m) = ar_m^{\alpha}\exp(-\alpha/\beta).$$ (1.20)

Because it has four adjustable constants, Equation 1.17 can be fitted to various aerosol models. The gamma distribution is usually employed to model haze, fog, and cloud particle size distributions.

Still another commonly used distribution is the lognormal distribution, which permits fitting the multimodal nature of the atmospheric aerosols. Aerosol size distributions are often modeled by the sum of two or three lognormal distributions, as follows:

$$n(r) = \frac{dN(r)}{dr} = \sum_{i=1}^{2,3}\frac{N_i}{r\sqrt{2\pi}\ln\sigma_i}\exp\left[-\frac{1}{2}\left(\frac{\ln(r/r_{mi})}{\ln\sigma_i}\right)^2\right]$$ (1.21)

where r_{mi} is the mode radius, σ_i is the standard deviation (measures the width of the distribution), and N_i is the total number of particles (in particles per cm³) that have r_{mi} as the mode radius. The mode radius depends on the aerosol production mechanism. The variation in the mode radius for different types of aerosol with relative humidity (RH) is given in Reference 34. Aerosols in the planetary boundary layer (up to 2–3 km) generally have a bimodal size distribution in the size range of optically effective particle radii; i.e., the accumulation particle mode is at around 0.03–0.5 μm in radius and the large particle mode at around 0.5–3 μm for volume size distribution. Maximal values of the individual modes change corresponding to different seasons or different atmospheric conditions.

An example of particulate size distribution behavior for different atmospheric conditions is illustrated in Figure 1.3. The measurements were performed in the semi-arid Negev desert (Israel) at a height of about 30 m above ground level using an optical particle counter

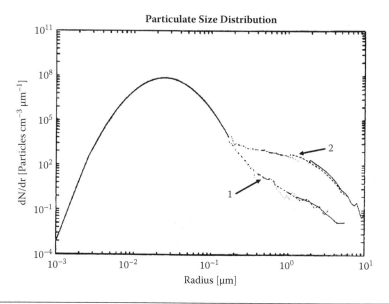

Figure 1.3   Particulate size distribution as a function of particle radius for different atmospheric conditions: (1) $T \sim 18.5°C$, $RH \sim 62\%$, $V \sim 0.8$ m/s; (2) $T \sim 12°C$, $RH \sim 44\%$, $V \sim 9.2$ m/s (sandstorm).

(Particle Measuring System Inc.). Strong continental wind brings additional solid suspended particles, and the size distribution moves toward larger particles, especially for particulates with $r \geq 0.2$ μm.

The tropospheric aerosols above the boundary layer are assumed to have the same composition, but their size distribution is modified by eliminating the large particle component.

To utilize the extensive measurements, a series of aerosol models for different environmental conditions and seasons were constructed [35]. The models have been divided into four altitude regimes, as described in Section 1.2.1. Many experiments have been carried out over the last three decades to test and modify the present aerosol models. The new data available show that the distribution varies dramatically with altitude, often within meters. Large variations exist in the data from different locations [16].

The relationship between the size distribution of the aerosol particles and their spectral extinction, α_λ, and backscatter, β_λ, coefficients is expressed by the Fredholm integral equation as

$$\alpha_\lambda = \int \pi r^2 Q_{ext}(r, \lambda, m) N(r) dr. \tag{1.22}$$

$$\beta_\lambda = \int \pi r^2 Q_{\text{bsc}}(r, \lambda, m) N(r) dr \qquad (1.23)$$

Here, the aerosol extinction (α_λ) and backscatter (β_λ) coefficients are measured at various wavelengths; r is the particle radius; m is the complex refractive index of aerosol particles, which equals $m_{Re} - im_{Im}$ according to its real and imaginary components [21]; $Q_{\text{ext}}(r, \lambda, m)$ and $Q_{\text{bsc}}(r, \lambda, m)$ are the extinction and backscatter efficiency factors, respectively; and $N(r)$ is the concentration of particles (in $\text{cm}^{-3}\ \mu\text{m}^{-1}$) with radii between r and $r + dr$. The imaginary part of the aerosol refractive index determines the absorption.

The kernel functions $[Q_{\text{ext}}(r, \lambda, m), Q_{\text{bsc}}(r, \lambda, m)]$ depend on the material of the aerosol particles as expressed by their complex refractive index m. Assuming spherical particles, the aerosol extinction can be calculated using classical Mie scattering theory [21] from a knowledge of the size distribution, number density, and complex refractive index. However, every irregularity in particle size leads to strong changes.

It is often found that within the optical sub-range (~0.16–1.2 μm), the light-scattering coefficient, α_λ, and aerosol size distribution in the form of Equation 1.15 obey the following power law relationship [12]:

$$\alpha_\lambda = C\lambda^{-b}, \qquad (1.24)$$

where b is referred to as the Angstrom exponent and $b = v - 3$. Thus, if α_λ depends strongly on wavelength (large b), then the size distribution function (Equation 1.15) decreases with particle size. Measurements have shown that values of b tend to be higher for continental aerosols than for clean marine aerosols.

Multiple aerosol scattering induces loss of laser beam intensity (attenuation) after it has traversed the scatterers and reduces image contrast and beam irradiance at the target. Besides, light scattering causes different angles of arrival of photons at the image plane, which contributes to image blur. In satellite imagery, the aerosol blur is considered to be the primary source of atmospheric blur (the turbulence blur is usually neglected) and is commonly called the *adjacency effect* [21]. In optical communication, the dense aerosol/dust layers and clouds as part of the communication channel can cause signal power attenuation and temporal and spatial widening. This effect limits the maximum data rate and increases the bit error rate (BER). Thus, knowledge of the parameters that determine the optical properties of atmospheric aerosols (spectral extinction and size distribution) is essential for development of techniques for optical communication and imaging through the atmosphere, laser weaponry, remote sensing in particular from space, or the necessary correction of atmospheric effects in satellite imagery.

1.3 Hydrometeors

Hydrometeors are any water or ice particles that have formed in the atmosphere or at the Earth's surface as a result of condensation or sublimation. Water or ice particles blown from the ground into the atmosphere are also classed as hydrometeors. Some well-known hydrometeors are rain, fog, snow, clouds, hail, dew, rime, glaze, blowing snow, and blowing spray. Scattering by hydrometeors has an important effect on signal propagation.

1.3.1 Fog

Fog is a cloud of small water droplets near ground level and sufficiently dense to reduce horizontal visibility to less than 1000 m. Fog is formed by the condensation of water vapor on condensation nuclei that are always present in natural air. This can result as soon as the relative humidity of the air exceeds saturation by a fraction of 1%. In highly polluted air the nuclei may grow sufficiently to cause fog at humidities of 95% or less. Three processes can increase the relative humidity of the air: (1) cooling of the air by adiabatic expansion, (2) mixing two humid air streams of different temperatures, and (3) direct cooling of the air by radiation (namely, cosmic ray radiation). According to the physical processes of fog creation, different kinds of fogs are usually observed: advection, radiation, inversion, and frontal. (The reader is directed to References 4 and 7 for a more detailed explanation of fog creation.)

When the air becomes nearly saturated with water vapor (RH → 100%), fog can form, assuming sufficient condensation nuclei are present. The air can become saturated in two ways, either by mixing of air masses with different temperatures and/or humidities (advection fogs), or by the air cooling until the air temperature approaches the dew point temperature (radiation fogs).

Fog models, which describe the range of different types of fog, have been widely presented based on measured size distributions [34]. The modified gamma size distribution (Equation 1.17) was used to fit the data. The models represent heavy and moderate fog conditions. Developing fog can be characterized by droplet concentrations of 100–200 particles per cm^3 in the 1–10 μm radius range with mean radius of 2–4 μm. As the fog thickens, the droplet concentration decreases to less than 10 particles per cm^3 and the mean radius increases from 6 to 12 μm. Droplets less than 3 μm in radius are observed in fully developed fog. It is usually assumed that the refractive index of the fog corresponds to that of pure water. Natural fogs and low-level clouds are composed of spherical water droplets, the refractive properties of which have been fairly well documented in the spectral region of interest.

1.3.2 Rain

Rain is precipitation of liquid water drops with diameters greater than 0.5 mm. When the drops are smaller, the precipitation is usually called drizzle. The concentration of raindrops typically spreads from 100 to 1,000 m^{-3}. Raindrops seldom have diameters larger than 4 mm because, as they increase in size, they break up. The concentration generally decreases as diameters increase. Meteorologists classify rain according to its rate of fall. The hourly rates relate to light, moderate, and heavy rain, which correspond to dimensions less than 2.5 mm, between 2.8 and 7.6 mm, and more than 7.6 mm, respectively. Less than 250 mm and more than 1500 mm per year represent approximate extremes of rainfall for all of the continents (see details in References 1, 6, 8, and 10).

1.3.3 Clouds

Clouds have dimensions, shape, structure, and texture, which are influenced by the kind of air movements that result in their formation and growth, and by the properties of the cloud particles. In settled weather, clouds are well scattered and small and their horizontal and vertical dimensions are only a kilometer or two. In disturbed weather, they cover a large part of the sky, and individual clouds may tower as high as 10 km or more. Growing clouds are sustained by upward air currents, which may vary in strength from a few centimeters per second to several meters per second. Considerable growth of the cloud droplets, with falling speeds of only about 1 cm/s, leads to their fall through the cloud and reaching the ground as drizzle or rain. Four principal classes are recognized when clouds are classified according to the kind of air motions that produce them:

1. Layer clouds formed by the widespread regular ascent of air
2. Layer clouds formed by widespread irregular stirring or turbulence
3. Cumuliform clouds formed by penetrative convection
4. Orographic clouds formed by ascent of air over hills and mountains

How such kinds of clouds are created is a matter of meteorology. Interested readers can find information in References 1–8.

Several alternative mathematical formulations have been proposed for the probability distribution of sky cover, as an observer's view of the sky dome. Each of them uses the variable x ranging from 0 (for clear conditions) to 1.0 (for overcast).

1.3.3.1 First Cloud Cover Model The *beta distribution* is an early cloud model whose density function is given by the following expression, accounting for the well-known gamma functions, $\Gamma(a)$, $\Gamma(b)$, and $\Gamma(a+b)$ [4, 5, 8]:

$$f(x) = \frac{\Gamma(a+b)}{\Gamma(a) \cdot \Gamma(b)} x^{a-1}(1-x)^{b-1}, \quad 0 \leq x \leq 1, \quad a, \; b > 0 \quad (1.25)$$

In [4, 5], the pairs of values of the two parameters, a and b, are presented for 29 regions over the world, for the four midseason months, and two times of day.

1.3.3.2 Second Cloud Cover Model Bean and Dutton [7] have proposed a model, called the *S distribution*, which is the cumulative probability distribution function $F(x)$ of sky cover x, estimated as [4, 7]:

$$F(x) = 1 - (1 - x^{\alpha})^{\beta}, \quad 0 \leq x \leq 1, \quad \alpha, \beta > 0 \quad (1.26)$$

Pairs of values of the two parameters (α, β) have been determined to make the distribution $F(x)$ fit the data in the sky-cover summaries. The data presented in References 4 and 7 have been published for 23 stations around the world, for each of eight periods of the day in each month of the year. The best pair of values for the sky cover in January at noontime was found to be $\alpha = 0.1468$, $\beta = 0.1721$.

1.3.3.3 Third Cloud Cover Model This model, also called *model B*, is used for linear and areal coverage of a weather element and has been described in References 4 and 5. Like the other two models, it requires two parameters for a description of the probability distribution of cloud cover, defined as the area smaller than the floor space of the sky dome. The parameters in model B have physical meaning. One parameter, P_0, is the mean cloud cover as given in climatic summaries; it is taken to be the single-point probability of a cloud intercept when looking up from the ground. The second parameter, r, known as the scale distance, is the distance between two stations whose correlation coefficient of cloud cover is 0.99.

1.3.3.4 Ceiling Cloud Model This is one of the best models for ceiling height cumulative distributions [4, 7]:

$$F(h) = 1 - \left[1 + \left(\frac{h}{c} \right)^a \right]^{-b} , \quad a, b, c > 0 \qquad (1.27)$$

where h is the ceiling height and a, b, and c are parameters. Bean and Dutton [7] set the values for a, b, and c, which have been determined for each of eight periods of the day in each month of the year at 23 stations around the world to make the estimated distributions $\hat{F}(h)$, fit the data for 30 ceiling heights. The best pair of values for the sky cover in January at noontime was found to be $a = 1.1678$ and $b = 0.1927$ when $c = 0.305$ km.

1.3.4 Snow

Snow is the solid form of water that crystallizes in the atmosphere and, falling to the Earth, covers permanently or temporarily about 23% of the Earth's surface. Snow falls at sea level poleward of latitude

35° N and 35° S, though on the west coast of continents it generally falls only at higher latitudes. Close to the equator, snowfall occurs exclusively in mountain regions, at elevations of 4900 m or higher. The size and shape of the crystals depend mainly on the temperature and the amount of water vapor available as they develop. In colder and drier air, the particles remain smaller and compact. Frozen precipitation has been classified into seven forms of snow crystals and three types of particles: graupel (granular snow pellets), also called soft hail; sleet (partly frozen ice pellets); and hail (hard spheres of ice) (see details in References 7 and 9).

1.4 Atmospheric Turbulence

The temperature and humidity fluctuations combined with turbulent mixing by wind and convection induce random changes in the air density and in the field of atmospheric refractive index in the form of optical turbules (or eddies), called optical turbulence, which is one of the most significant parameters for optical wave propagation [36–44]. Random space-time redistribution of the refractive index causes a variety of effects on an optical wave related to its temporal irradiance fluctuations (scintillations) and phase fluctuations. A statistical approach is usually used to describe both atmospheric turbulence and its various effects on optical or IR systems.

The aim of this section is to provide the theoretical background needed for the statistical description of optical turbulence and its implications. We review the atmospheric phenomenology giving rise to turbulent effects and discuss the various models available to describe and quantify optical turbulence in the Earth's atmosphere.

1.4.1 Energy Cascade Theory

In the earliest study of turbulent flow, Reynolds used similarity theory to define a non-dimensional quantity Re = VL/v, called the Reynolds number, where V and L are the characteristic velocity and size of the flow, respectively, and v is the kinematic viscosity (in m²/s). The transition from laminar to turbulent motion takes place at a critical Reynolds number, above which the motion is considered turbulent.

Because the kinematic viscosity, v, of air is of the order 10^{-5} m^{2}s^{-1}, air motion is considered highly turbulent in the surface layer and in the free atmosphere (boundary layer and troposphere), where the Reynolds numbers are on the order Re ~ 10^5 [17, 19, 20].

Richardson [44] first developed a theory of the turbulent energy redistribution in the atmosphere — the energy cascade theory. It was noticed that smaller scale motions originated as a result of the instability of larger ones. A cascade process, in which eddies of the largest size are broken into smaller and smaller ones, continues down to scales in which the dissipation mechanism turns the kinetic energy of motion into heat.

Kolmogorov [45] introduced a hypothesis stating that during the cascade process the direct influence of larger eddies is lost and smaller eddies tend to have independent properties, universal for all types of turbulent flows. Following Kolmogorov, the energy cascade process consists of an energy input region, inertial sub-range, and energy dissipation region.

At a large characteristic scale or eddy, a portion of kinetic energy in the atmosphere is converted into turbulent energy. When the characteristic scale reaches a specified outer scale size, L_0, the energy begins a cascade that forms a continuum of eddy size for the energy transfer from a macroscale, L_0, to a microscale, l_0, called the inner turbulence scale. The scale sizes, l, bounded above by L_0 and below by l_0, form the inertial sub-range.

Kolmogorov [45] proposed that in the inertial sub-range, where $L_0 > l > l_0$, turbulent motions are both homogeneous and isotropic and energy may be transferred from eddy to eddy without loss; i.e., the amount of energy that is being injected into the largest structure must be equal to the energy that is dissipated as heat.

The term *homogeneous* is analogous to stationarity and implies that the statistical characteristics of the turbulent flow are independent of position within the flow field. The term *isotropic* requires that the second and higher order statistical moments depend only on the distance between any two points in the field.

The inertial sub-range is dominated by inertial forces, and the average properties of the turbulent flow are determined only by the average dissipation rate, ε (in m^2/s^3), of the turbulent kinetic energy. When the size of a decaying eddy reaches l_0, the energy is dissipated as heat through viscosity processes. It also has been hypothesized that the motion associated with the small-scale structure l_0 is uniquely determined by the kinematic viscosity, v, and ε, where $l_0 \sim \eta = (v^3/\varepsilon)^{1/4}$ is the Kolmogorov microscale [17, 19, 45]. The Kolmogorov microscale defines the eddy size dissipating the kinetic energy.

The fundamental characterization of atmospheric turbulence was developed by Kolmogorov in terms of the velocity field fluctuations [19, 45]. It was assumed that the velocity fluctuations can be represented by a locally homogeneous and isotropic random field for scales less than the large eddies or the energy source, implying that the second and higher order statistical moments of the turbulence depend only on the distance between any two points in the structure. In general, turbulent flow in the atmosphere is neither homogeneous nor isotropic. However, it can be considered locally homogeneous and isotropic in small sub-regions of the atmosphere.

Using dimensional analysis, Kolmogorov showed that the structure function of the velocity field in the inertial sub-range satisfies the universal 2/3 power law as

$$D_V(r) = <[V(\mathbf{r}_1 + \mathbf{r}) - V(\mathbf{r}_1)]^2> = C_V^2 \, r^{2/3}, \quad l_0 < r < L_0 \quad (1.28)$$

where the angle brackets denote average (time or ensemble, assuming ergodicity), $V(\mathbf{r}_1)$ is the turbulent component of velocity vector at point $\mathbf{r}_1$, $r = |\mathbf{r}|$ is the distance between the two observation points, and C_V^2 is the velocity structure constant — a measure of the total amount of energy in the turbulence. The structure constant is related to the average energy dissipation rate, ε, by $C_V^2 = 2\varepsilon^{2/3}$ [45]. The velocity field inner scale, l_0, is on the order of the Kolmogorov microscale, η, and is given by $l_0 = 12.8\eta$ [19]. Note that l_0 increases with kinematic viscosity, which increases with altitude. The inverse dependence of inner scale on the average rate of dissipation, ε, shows that strong turbulence has smaller inner scales and weak turbulence has larger inner scales. The inner scale is typically on the order of 1 to 10 mm near the ground and can be on the order of centimeters or more in the troposphere and stratosphere [17, 19].

The outer scale, L_0, is proportional to $\varepsilon^{1/2}$, and it increases and decreases directly with the strength of turbulence [17, 19]. The outer scale, L_0, can be scaled with height, h, above ground in the surface layer up to ~100 m according to $L_0 = 0.4h$ [19].

The behavior of the velocity field structure function at small-scale sizes ($r < l_0$) varies with the square of separation distance, as follows [45]:

$$D_V(r) = C_V^2 \, l_0^{-4/3} \, r^2, \quad 0 < r < l_0 \qquad (1.29)$$

Because the random field of velocity fluctuations is basically nonisotropic for scale sizes larger than the outer scale, L_0, no general description of the structure function can be predicted for $r > L_0$.

The validity of the 2/3 power law for the structure function has been established over a wide range of experiments [46, 47]. The region of greatest interest is the inertial sub-range, where the 2/3 power law contains all information on turbulence of practical importance.

As mentioned, the outer scale, L_0, denotes the scale size below which turbulence properties are independent of the flow and generally nonisotropic. The source of energy at large scales is either wind shear or convection. Scale sizes smaller than the inner scale belong to the viscous dissipation range. In this regime, the turbulent eddies disappear and the energy is dissipated as heat.

The turbulent process is shown schematically in Figure 1.4, with an energy input region, inertial sub-range, and energy dissipation region.

The physical origin of the optical effects of atmospheric turbulence is random index-of-refraction fluctuations, also called optical turbulence. For optical wave propagation, refractive index fluctuations are caused primarily by fluctuations in temperature. (Variations in humidity and pressure can be neglected.)

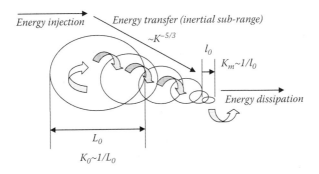

Figure 1.4 Schematic description of turbulent cascade process. K is the spatial wavenumber $K \sim 1/L$; L is the eddy size.

The statistical description of the random field of turbulence-induced fluctuations in the atmospheric refractive index is similar to that for the related field-of-velocity fluctuations.

The concept of a conservative passive additive (passive scalar) allowed Obukhov [48] to relate the velocity structure function to the structure function for potential temperature and then to the structure function for the variations in the refractive index, which satisfy the relationship

$$D_n(r) = <[n(\mathbf{r}_1 + \mathbf{r}) - n(\mathbf{r}_1)]^2> = C_n^2 \, r^{2/3}, \quad l_0 < r < L_0 \qquad (1.30)$$

where $r = |\mathbf{r}_1 - \mathbf{r}_2|$, C_n^2 is the refractive index structure constant (in $m^{-2/3}$), and is a measure of magnitude of fluctuations in the index of refraction. It characterizes the strength of the refractive turbulence. It is the critical parameter for describing optical turbulence and is often used synonymously with optical turbulence.

Generally speaking, the refractive index structure function describes the behavior of correlations of passive scalar field fluctuations between two given points separated by a distance r. The quantity L_0 is a measure of the largest distances over which fluctuations in the index of refraction are correlated, whereas l_0 is a measure of the smallest correlation distances. As was described, the correlation distances L_0 and l_0 are usually referred to as the outer and inner scale size of the turbulent eddies, respectively.

The inner scale of refractive index fluctuations is related to the Kolmogorov microscale by $l_0 = 7.4\eta = 7.4(v^3/\varepsilon)^{1/4}$ [46, 49]. In the dissipative interval, where $r < l_0$, the refractive index structure function corresponds to behavior in the form of Equation 1.29. Values of refractive index C_n^2 near ground typically range from about 10^{-16} $m^{-2/3}$ or less for "weak turbulence" up to 10^{-13} $m^{-2/3}$ or more when the turbulence is "strong."

Over short time intervals at a fixed propagation distance and constant height above uniform surface, it may often be assumed that C_n^2 is essentially constant. However, for vertical and slant-path propagation, C_n^2 varies as a function of height above ground.

Fluctuations in the index of refraction are related to corresponding temperature, humidity, and pressure fluctuations. Because pressure fluctuations are usually negligible, the refractive index fluctuations associated with the visible and near-IR region of the spectrum are due primarily to random temperature fluctuations. The wavelength dependence is small for optical frequencies. Humidity effects are typically neglected over land, because humidity affects the value of the refractive index by less than 1%. Humidity fluctuations contribute only in the far-IR region. However, for high water humidity and temperature fluctuations, humidity covariance can be significant [21].

The values of C_n^2 are closely related to the temperature structure parameter by the following equation [19, 20]:

$$C_n^2 = (79 \cdot 10^{-6} \, P/T^2)^2 \, C_T^2 \tag{1.31}$$

This relation is valid for propagation of visible or near-IR radiation over land. From Equation 1.31 it follows that structure functions of refractive index and of temperature random fluctuations differ only by constant coefficient.

1.4.2 Spectral Characteristics

The wavenumber power spectrum of refractive index fluctuations in the atmosphere has important consequences on a number of problems involving the propagation and scattering of electromagnetic waves.

The 2/3 power law behavior of the structure function in the inertial sub-range is equivalent to the power spectrum in three dimensions (also called the Kolmogorov turbulence spectrum), given by [45–47]

$$\Phi_V(K) = 0.033 C_V^2 \, K^{-11/3}, \quad 2\pi/L_0 < K < 2\pi/l_0 \tag{1.32}$$

where K is the spatial frequency or wavenumber (in rad m^{-1}). The three-dimensional power spectrum with $-11/3$ power law is related to the one-dimensional (1D) spectrum with $-5/3$ power law by

$$\Phi_V(K) = -\frac{1}{2\pi K} \frac{d\Phi_V^{1D}(K)}{dK} \tag{1.33}$$

where $\Phi_V{}^{1D}(K)$ is the 1D power spectrum. The spectrum measures the distribution of the variance of a variable over scale sizes or periods. If the variable is a velocity component, the spectrum also describes the distribution of kinetic energy over spatial periods.

Using the relation between the structure function and the covariance, and the Wiener-Khinchin theorem, the relation between the structure function and the power spectrum is given by [20, 46]

$$D_V(r) = 8\pi \int_0^\infty K^2 \Phi_V(K) \left(1 - \frac{\sin(rK)}{rK} \right) dK \qquad (1.34)$$

The velocity power spectrum is not of interest when the optical properties of turbulence need to be characterized. In this case Obukhov's law [48] is adopted, which states that fluctuations in passive scalar quantities (e.g., temperature field, refractive index, etc.) associated with turbulent flow inherit the same power spectrum of fluctuations in flow, which has the form of Equation 1.32, under the assumption that the turbulent field is locally homogeneous and isotropic. Therefore, if $\Phi_S(k)$ describes the power spectrum of a given passive scalar, then $\Phi_S(k) \sim C_S^2 \cdot k^{-5/3}$, where $C_S^2 = const \cdot \varepsilon_T/\varepsilon^{1/3}$ is the structure constant derived by Obukhov [48]. Here, ε_T is the heat flux intensity over the spectrum (thermal dissipation rate).

Thus, based on Obukhov's law and the 2/3 power law (Equation 1.30) for the structure function, the associated three-dimensional spectrum for the refractive index fluctuations over the inertial sub-range is defined by the Kolmogorov spectrum in form of Equation 1.32 by

$$\Phi_n(K,z) = 0.033\, C_n^2(z) K^{-11/3}, \quad 2\pi/L_0 < K < 2\pi/l_0 \qquad (1.35)$$

where $\Phi_n(K,z)$ is the power spectral density of refractive index fluctuations as a function of position along the optical path z, and K is the spatial wavenumber. Equation 1.35 is known as the Obukhov-Kolmogorov power spectrum. The model is valid only for the inertial sub-range, although it is often extended over all wavenumbers by assuming the inner scale is zero and the outer scale is infinite. It is usually assumed that the inertial sub-range determines the optical properties of the turbulent atmosphere.

Other spectrum models have been proposed for making calculations when inner scale and/or outer scale effects cannot be ignored. The following isotropic forms of spectra, which take into account deviations from a power law in the region-of-turbulence outer and inner scales, can be used in calculations [20]:

The modified von Karman spectrum

$$\Phi_n(K) = 0.033C_n^2 \left(K^2 + K_0^2\right)^{-11/6} \exp\left[-K^2/K_m^2\right] \quad (1.36)$$

An exponential spectrum

$$\Phi_n(K) = 0.033C_n^2 K^{-11/3}\{1 - \exp[-K^2/K_0^2]\} \quad (1.37)$$

where $K_m = 5.92/l_0$ and $K_0 = C_0/L_0$, C_0 being the scaling constant for the outer scale parameter, which is typically chosen in the range of $1 \le C_0 \le 8\pi$.

Note that these forms of the turbulence spectrum are used for computational reasons only and are not based on physical models.

Hill [49, 50] performed a hydrodynamic analysis that led to a numerical spectral model with the small rise (or bump) at high wavenumbers near $K_m \sim 1/l_0$ that fit the experimental data of the temperature spectrum [50–52]. The bump in temperature spectrum also induces a corresponding spectral bump in the spectrum of the refractive index fluctuations, which can have important consequences on a number of statistical quantities important in problems involving optical wave propagation.

More recently, Andrews [20, 53] developed an approximation to the Hill spectrum — *modified atmospheric spectrum* — given by

$$\Phi_n(K) = 0.033C_n^2[1 + a_1(K/K_l) - a_2(K/K_l)^{7/6}]\frac{\exp\left[-K^2/K_l^2\right]}{\left(K^2 + K_0^2\right)^{11/6}}$$

$$(1.38)$$

where $a_1 = 1.802$, $a_2 = 0.254$, and $K_l = 3.3/l_0$.

Note that Equation 1.38 is similar to the functional form of Equation 1.36, except for the terms within the brackets that characterize the high wavenumber spectral bump.

In the simple Kolmogorov model of turbulence, the atmosphere is usually described as a single turbulent layer in which the variations of the refractive index with temperature and pressure induce both phase and amplitude fluctuations of the propagating wavefront. In addition, it is usually assumed that the time scale of temporal changes in the atmospheric layer of wave propagation is much smaller than the time it takes the wind to blow the turbulence over the receiver aperture (Taylor's hypothesis of frozen turbulence). The spatial and temporal properties of this single layer are thus linked by the wind speed, V, of the layer as $\omega = K \cdot V$, where ω is the frequency in radians per seconds.

1.4.3 C_n^2 Altitude Distribution

The refractive index structure coefficient, C_n^2, is the single most important parameter appearing in almost all equations that describe wavefront disturbances and image degradations caused by refractive turbulence. It can often be assumed to be fairly constant along a horizontal propagation path over a uniform surface. For slant-path propagation, the structure constant may vary because of its altitude dependence. In general, C_n^2 decreases with increasing height above the ground.

The C_n^2 profile is a function of the observation site, and its magnitude and functional behavior may vary strongly from one observation point to another. Modeling efforts predicting the atmospheric refractive turbulence at different altitude elevations are based on limited databases and do not work properly at different locations. The existing models of C_n^2 profiles can be divided into nonparametric and parametric ones. The nonparametric models represent average profiles (without stratification) and, in most cases, are site dependent. The parametric models have been developed in an attempt to incorporate the dependence on site and meteorological parameters and to introduce the stratification into turbulence profile.

The nonparametric C_n^2 profile models for the atmospheric boundary layer (called the Kaimal family models) [54, 55] apply only in convection-dominated conditions, and C_n^2 reference measurements need to be made. The atmospheric boundary layer parametric model is presented in Reference 21.

The free-atmosphere nonparametric models are the SLC (submarine laser communication) daytime and nighttime models [56] and the AFGL AMOS [57] and AFGL CLEAR I models [58]. These models are site dependent. Data for the SLC and AMOS models is gathered in a marine and mountain subtropical atmosphere. The AFGL CLEAR I model was developed for the New Mexico desert in the summer. The parametric free-atmosphere models include dependencies on winds and meteorology. One of the most used models is the Hufnagel-Valley (H-V) model [20, 59]. This is a one-parameter model that is determined from the upper altitude winds. However, this is a midlatitude model (a subtropical atmosphere), and its performance can be poor for other sites.

Another parametric model is the NOAA/Van Zandt, which incorporates the structure of wind shears and the potential temperature [60, 61]. An advantage of the NOAA/Van Zandt model is that it is a layered model, based on atmospheric physics of gravity wave generation. Other parametric models can be found in References 58, 62, and 63.

In the Middle East, although C_n^2 predictive models were developed [21], all were for elevations close to the ground and have application for horizontal propagation. Recently, the Middle East model of C_n^2 vertical profile has been developed on the basis of imaging LIDAR (light detection and ranging) measurements [64]. The model shows various layers in the vertical turbulence profile. There is a strong turbulence layer near the tropopause at a height of 8–9.5 km. This is confirmation of the theories concerning increased thermal turbulence near the tropopause. Other turbulence increases are in layers just above the atmospheric boundary layer, at elevations of 2.5–3 km and 5.5–6 km.

Distinctly layered turbulence structure has been confirmed in many measurements performed in the past employing passive and active methods and *in situ* observations [65–68]. The results indicate that the layer thicknesses can be different and change from 20 to 200 m. The strongest layers were typically present in the lower troposphere below 5 km and in the tropopause region 9 km and above. Such turbulence layers were observed also from measurements performed with scintillation detection and ranging (SCIDAR). The results indicate that the turbulence is concentrated typically in two or three layers [69, 70]. The physical mechanism generating such distribution was studied in Reference 71. The height of turbulent layers can differ with location and climate.

Some of the models are reviewed in this section, but a more comprehensive treatment can be found in References 58 and 72. The H-V model of the C_n^2 profile, which is most commonly used by the technical community in studies of vertical or slant-path optical propagation through the atmosphere, is described by

$$C_n^2(h) = a_1(V/27)^2 \, (h/s_1)^{10} \exp[-h/s_1] + a_2 \exp[-h/s_2] + C_n^2(0) \exp[-h/s_3],$$

(1.39)

with $a_1 = 5.94 \times 10^{-23}$, $a_2 = 2.7 \times 10^{-16}$, $s_1 = 1{,}000$ m, $s_2 = 1{,}500$ m, and $s_3 = 100$ m, and where h is altitude (m), V is the RMS wind speed (m/s), which controls high-altitude turbulence in the model, and $C_n^2(0)$ is the turbulence strength at the ground level (m$^{-2/3}$). The strong turbulence layer in the low stratospheric region has been introduced in the H-V model as an empirical parameter. The Middle East height-dependent model of C_n^2 is given by

$$C_n^2(h) = a_1 \exp\left[-\frac{h}{s_1}\right] + a_2 \exp\left[-\frac{h}{s_2}\right] + a_3 \exp\left[-\frac{h}{s_3}\right] + a_4 h^{10} \exp\left[-\frac{h}{s_4}\right]$$

$$+ a_5 \exp\left[-\frac{(h-H_1)^2}{2d_1^2}\right] + a_6 \exp\left[-\frac{(h-H_2)^2}{2d_2^2}\right]$$

$$+ a_7 \exp\left[-\frac{(h-H_3)^2}{2d_3^2}\right],$$

(1.40)

where h is the altitude elevation in meters; $a_1 = 8 \times 10^{-14}$; $a_2 = 10^{-15}$; $a_3 = 5.5 \times 10^{-17}$; $a_4 = 4.12 \times 10^{-53}$; $s_1 = 200$; $s_2 = 1{,}250$; $s_3 = 5{,}700$; $s_4 = 1{,}000$; $a_5 = 4.5 \times 10^{-16}$; $a_6 = 1.3 \times 10^{-16}$; $a_7 = 6.4 \times 10^{-16}$; $H_1 = 2{,}500$; $H_2 = 5{,}300$; $H_3 = 8{,}500$; $d_1 = 100$; $d_2 = 130$; and $d_3 = 130$.

The Middle East turbulence model form is similar to a generalized Hufnagel-Valley model [73], but with the additional layers. The H-V and the Middle East models are shown in Figure 1.5. The HV-21 model (dashed line) was used with the following parameters: $V = 21$ m/s and $C_n^2(0) = 1.7 \times 10^{-14}$ m$^{-2/3}$.

1.4.4 L_0 Altitude Distribution

Information about behavior of the outer scale as a function of altitude has not been precisely determined and is still questionable. The measurements based on SCIDAR observations suggest that the outer scale changes with altitude according to the empirical

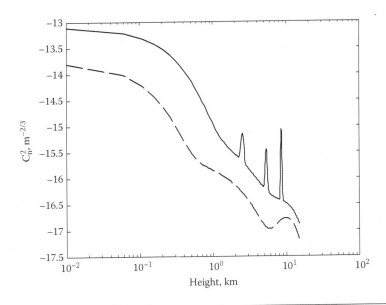

Figure 1.5 H-V 5/7 (dashed) and Middle East (solid) turbulence strength (C_n^2) vertical profile models.

formula [74]

$$L_0(z) = \frac{A_n}{1 + \left(\frac{z - B_n}{2500}\right)^2} \tag{1.41}$$

where $n = 1$, $A_1 = 4$ m and $B_1 = 8500$ m or $n = 2$, $A_2 = 5$ m and $B_2 = 7500$ m.

Recent measurements of turbulence parameters in the upper atmosphere led to outer scale values in the range of $5 < L_0 < 190$ m [20, 74–79]. The L_0 median values, ranging from 24 to 31 m, were obtained with a generalized seeing monitor (GSM) at different sites [71, 72].

$L_0(z)$ can be modeled in the form

$$L_0(z) = 0.4z \quad \text{for } z \leq 25 \text{ m} \tag{1.42a}$$

$$L_0(z) = \frac{45}{1 + \left(\frac{z - 8000}{4500}\right)^2} \quad \text{for } z > 25 \text{ m} \tag{1.42b}$$

This proposed model is based on analysis of different measurement results [20, 74–81].

1.4.5 Non-Kolmogorov Turbulence

Assumption of Kolmogorov's turbulence spectrum can prove problematic in the description of some characteristics of electromagnetic (EM) wave propagation in the atmosphere, because the atmospheric turbulence properties can differ considerably from those described by classic Kolmogorov theory and, in the case of a passive scalar, by Obukhov-Kolmogorov's theory.

To date, all calculations relating to EM wave propagation in the Earth's atmosphere are based on the Obukhov-Kolmogorov (O-K) model of the refractive index fluctuation spectrum. However, as mentioned by Kerr [82], "the degree of universality of the Kolmogorov spectrum is a matter of controversy."

Further support for deviations from the O-K model has been obtained in Earth-space microwave propagation experiments as well as in optical and *in situ* measurements [46, 82–86]. In the past decade, additional experimental evidence has shown significant deviations from the O-K model in certain portions of the atmosphere [85–92]. As was observed, the power spectrum of turbulence in the free troposphere and stratosphere may exhibit non-Obukhov-Kolmogorov (N-O-K) properties. This means that for elevations above the boundary layer, models based on Kolmogorov turbulence are not applicable.

The experimental results point toward a more general equation, given by

$$\Phi_n(K) = const\ C_n^2\ K^{-\alpha}, \tag{1.43}$$

where $\alpha = 11/3$ (or $\alpha = 5/3$ for the 1D spectrum) in the case of O-K turbulence.

Depending on atmospheric conditions, the power spectrum, $\Phi_n(K)$, can acquire various forms, as presented in Figure 1.6, and α can exhibit a broad range of values different from 5/3 for O-K turbulence or the 5/3 value can be part of a more broad spectral distribution [92, 93].

As demonstrated previously by Gibson and others [95, 96], when the Prandtl number $Pr = v/B_d$, which determines the smallest scale fluctuations of scalar fields mixed by turbulence (where B_d is the molecular diffusion coefficient), becomes greater than unity, a short zone appears in the vicinity of the diffusion interval, where the 1D

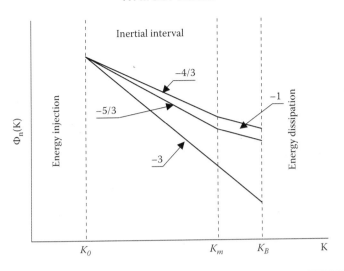

Figure 1.6 Schematic dependence of the 1D power spectrum vs. wave vector for different turbulence models. Here K_0 corresponds to the boundary between large scales and the inertial interval, K_m is between the inertial interval and Batchelor's interval [94], and K_B is between the Batchelor and diffusion intervals. The numbers represent a in Equation 1.43. (From A. Zilberman, E. Golbraikh, and N. S. Kopeika, "Lidar studies of aerosols and non-Kolmogorov turbulence in the Mediterranean troposphere," *Proc. SPIE*, vol. 5987, pp. 15–26, 2005. With Permission.)

turbulent fluctuations spectrum has a slope $\alpha = 1$ (Batchelor spectrum [94]). With growing average temperature and, consequently, growing B_d value, the boundaries of this interval become blurred and disappear at Pr < 1. Depending on the Pr magnitude, the influence of this interval on the turbulent spectrum, and hence on its various convolutions in atmospheric optic problems, can grow.

With increasing scale sizes of passive scalar turbulent fluctuations, a viscous-convective zone with $\alpha = 1$ is passed to the inertial interval. The magnitude of the wave scale, K_m, separating these intervals can be equated, to a first approximation, to the magnitude of the wave vector determining the boundary of the dissipative interval of the turbulent velocity field (i.e., $K_m \sim 1/l_0 = [\varepsilon/v^3]^{1/4}$). At the transition to the inertial interval, it is assumed as a rule that the 1D spectrum has slope $-5/3$ (O-K turbulence).

Another type of turbulence is as widespread as Kolmogorov's; namely, *helical* turbulence [97]. There are two limiting cases, where the spectral density $E(K)$ of the energy of turbulent fluctuations of velocity field in the inertial interval depends either on energy flux, ε, or on *helicity* flux, ε_η [97]. In the first case, the turbulence energy spectrum is $E(K) \sim \varepsilon^{2/3}K^{-5/3}$ (Kolmogorov case), and in the second, $E(K) \sim \varepsilon_{\eta}^{2/3}K^{-7/3}$ (helical turbulence), where K is the wave vector. Such spectra are observed in laboratory magnetohydrodynamic flows and in atmospheric experiments at various altitudes.

Further studies of mean helicity effect on turbulent fluctuations have shown that both the energy spectrum of the turbulent velocity field and the spectrum of passive scalar fluctuations have the form of Equation 1.43, with the exponent $\alpha = 7/3$ for the velocity field and $\alpha = 4/3$ for the passive scalar field (for the 1D spectrum) [97, 98]. If the helicity is strong enough, then the spectrum can change, and Equation 1.43 takes the form [93]

$$\Phi_n^{1D}(K) = C_n^2 K^{-4/3} \tag{1.44}$$

where $C_n^2 = const\ \varepsilon_T/\varepsilon_{\eta}$, where ε_{η} is the helicity flux over the spectrum, and Φ_n^{1D} is the 1D spectrum.

The power spectrum (Equation 1.44) is related to the corresponding structure function, $D(r)$. The velocity field structure function in the case of helical turbulence corresponds to $D_V(r) \sim r^{4/3}$. The structure function of the passive scalar field is changed in comparison with the O-K case to $D_{ps}(r) \sim r^{1/3}$. Thus, for helical turbulence, characteristics of the propagating EM wave should also vary.

Table 1.1 provides the spectral index values relating to velocity field and passive scalar field for O-K and helical turbulence.

An absolutely different kind of spectrum arises in the inertial interval of a passive scalar spectrum when the influence of the velocity field becomes negligible. In this case Pr < 1, and a situation can arise [96] in which the properties of passive scalar fluctuations in an inertial interval no longer depend on the velocity field behavior, the spectrum becomes steeper than in the O-K case, and α acquires the value of 3 ($\alpha = 5$ for 3D spectrum). As shown elsewhere [99], when the passive scalar

Table 1.1 Power Law Exponents for Different Turbulence Models (Inertial Interval)

	VELOCITY FIELD		PASSIVE SCALAR FIELD	
	KOLMOGOROV	HELICAL	KOLMOGOROV	HELICAL
$D(r) \sim r^p$	$p = 2/3$	$p = 4/3$	$p = 2/3$	$p = 1/3$
$\Phi^{1D}(K) \sim K^{-\alpha}$	$\alpha = 5/3$	$\alpha = 7/3$	$\alpha = 5/3$	$\alpha = 4/3$

Note: $\Phi^{1D}(K)$ is the 1D spectrum.

distribution becomes anisotropic, the spectrum with $\alpha = 3$ is extended over the entire inertial interval, as shown in Figure 1.6. (In this case, the turbulent velocity field can remain uniform and isotropic.)

Such spectra of atmospheric inhomogeneity fluctuations are characteristic of the stratosphere [85, 90]. One-dimensional temperature power spectra measured in the stratosphere by different techniques (remotely and *in situ*) and predicted by the theory of saturated gravity waves have a slope of −3 instead of −5/3. The existence of the spectrum of passive scalar fluctuations with $\alpha \approx 3$ in the Earth's troposphere was recently confirmed by experiments of lidar sounding [92, 100]. It was found that spectra of similar type are observed in the troposphere starting from altitudes above ~7–8 km.

It should be noted that N-O-K spectra can exist in the presence of Kolmogorov's turbulent velocity field. Thus, when the non-Kolmogorov characteristics of turbulence are mentioned, they refer to the behavior of the passive scalar spectrum and not the velocity field for which the Kolmogorov theory was developed.

1.4.6 Generalized Power Spectrum

The three-dimensional (3D) power spectrum of refractive index fluctuations in terms of the arbitrary power law can be defined as

$$\Phi_n(K, \alpha, z) = A(\alpha) \cdot \beta(z) \cdot K^{-\alpha} \tag{1.45}$$

where K is the spatial wavenumber, $\beta(z)$ is the general refractive index structure constant (similar to C_n^2) and has units of $m^{3-\alpha}$, and $A(\alpha)$ is a constant that maintains consistency between the index structure function and its power spectrum. The solution for $A(\alpha)$ has the form [84]

$$A(\alpha) = \frac{\Gamma(\alpha-1)}{4\pi^2} \sin\left[(\alpha-3)\frac{\pi}{2}\right] \quad 3 < \alpha < 5, \tag{1.46}$$

where $\Gamma(x)$ is Euler's gamma function.

Equation 1.45 reduces to the Kolmogorov power spectrum if $\alpha = 11/3$. In this case $A(11/3) = 0.033$. For helical turbulence $A(10/3) = 0.015$, and for non-Kolmogorov anisotropic turbulence $A(\alpha \to 5) \sim 0.0024$. For the intermediate case, if slope is between 10/3 and 5, $A(\alpha)$ can have a maximum value equal to ~0.061.

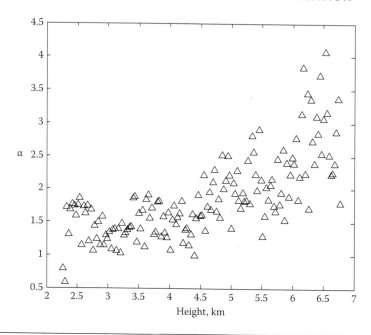

Figure 1.7 Changes in spectral exponent α (1D spectrum) with altitude [100]; 30 m altitude resolution.

The power spectrum corresponds to an arbitrary structure function:

$$D_n(r,z) = \beta(z)\, r^p, \qquad \text{with } p = \alpha - 3 \qquad (1.47)$$

The spectral exponent α is not a universal turbulence parameter in the atmosphere and it depends on many factors, as obtained in different works (see References 86 and 93 and references therein).

It has been found that spectral behavior of the refractive index fluctuations changes with altitude, which makes the use of universal models such as the O-K model quite inappropriate. An example of such behavior is shown in Figure 1.7 [100]. The spectral exponent α grows with altitude and varies from values close to 4/3 or 5/3 (helical/O-K spectrum [85]) to 3 (independent anisotropic spectrum [99]). Thus, altitude dependence appears in the structure function not only regarding $C_n^2(z)$, but also the spectral index α.

Based on the results of experimental and theoretical works [85–89, 92, 101–103], it may be assumed that the troposphere and lower stratosphere are composed of three main turbulent layers where spectral index α is a constant inside each: the first layer corresponds to the boundary layer (up to ~2–3 km) with O-K turbulence ($\alpha = 11/3$); the second one corresponds to the free troposphere (up to ~8–10 km), where N-O-K turbulence has slope –10/3 (helical N-O-K spectrum); and the third (>10 km) where the N-O-K turbulence spectrum has a slope of –5 [85]. Thus, the generalized form of the power spectrum of refractive index fluctuations changes as a function of altitude with $\alpha \equiv \alpha(z)$ in Equation 1.45.

It is noteworthy that a multilayer model of the atmosphere has been used practically always for solving problems of EM radiation transfer in the atmosphere, with the structure constant, C_n^2, varying with altitude. However, its strong change (by orders of magnitude) with altitude is probably connected with changes in turbulent fluctuation spectrum with altitude.

References

1. H. R. Pruppacher and R. L. Pitter, "A semi-empirical determination of the shape of cloud and rain drops," *J. Atmos. Sci.*, vol. 28, pp. 86–94, 1971.
2. A. Slingo, "A GSM parametrization for the shortwave radiative properties of water clouds," *J. Atmos. Sci.*, vol. 46, pp. 1419–1427, 1989.
3. M. D. Chou, "Parametrizations for cloud overlapping and shortwave single scattering properties for use in general circulation and cloud ensemble models," *J. Climate*, vol. 11, pp. 202–214, 1998.
4. International Telecommunication Union, "Attenuation due to clouds and fog," ITU-R Recommendation P.840-2, Geneva, 1997.
5. K. N. Liou, *Radiation and Cloud Processes in the Atmosphere*, Oxford University Press, Oxford, England, 1992.
6. S. R. Saunders, *Antennas and Propagation for Wireless Communication Systems*, John Wiley & Sons, New York, 1999.
7. B. R. Bean and E. J. Dutton, *Radio Meteorology*, Dover, New York, 1966.
8. N. Blaunstein and C. Christodoulou, *Radio Propagation and Adaptive Antennas for Wireless Communication Links: Terrestrial, Atmospheric, and Ionospheric*, Wiley InterScience, Hoboken, NJ, 2007.
9. P. S. Ray, "Broadband complex refractive indices of ice and water," *Appl. Opt.*, vol. 11, pp. 1836–1844, 1972.
10. International Telecommunication Union, "Specific attenuation model for rain for use in prediction methods," ITU-R Recommendation P.838, Geneva, 1992.
11. S. Twomey, *Atmospheric Aerosols*, Elsevier, Amsterdam, 1977.
12. E. J. McCartney, *Optics of the Atmosphere: Scattering by Molecules and Particles*, John Wiley & Sons, New York, 1976.
13. K. Y. Whitby, "The physical characteristics of sulfur aerosols," *Atmos. Environ.*, vol. 12, pp. 135–159, 1978.

14. S. K. Friedlander, *Smoke, Dust and Haze,* John Wiley & Sons, New York, 1977.

15. J. H. Seinfeld, *Atmospheric Chemistry and Physics of Air Pollution,* John Wiley & Sons, New York, 1986.

16. G. A. d'Almeida, P. Koepke, and E. P. Shettle, *Atmospheric Aerosols, Global Climatology and Radiative Characteristics,* A. Deepak Publishing, Hampton, VA, 1991.

17. A. Ishimaru, *Wave Propagation and Scattering in Random Media,* Academic Press, New York, 1978.

18. S. M. Rytov, Yu. A. Kravtsov, and V. I. Tatarskii, *Principles of Statistical Radiophysics,* Springer, Berlin, 1988.

19. V. I. Tatarskii, *Wave Propagation in a Turbulent Medium,* McGraw-Hill, New York, 1961.

20. L. C. Andrews and R. L. Phillips, *Laser Beam Propagation through Random Media,* 2nd ed., SPIE Press, Bellingham, WA, 2005.

21. N. S. Kopeika, *A System Engineering Approach to Imaging,* SPIE Press, Bellingham, WA, 1998.

22. *U.S. Standard Atmosphere,* U.S. GPO, Washington, DC, 1976.

23. V. A. Kovalev and W. E. Eichinger, *Elastic Lidar: Theory, Practice, and Analysis Methods,* John Wiley & Sons, Hoboken, NJ, 2004.

24. C. E. Junge, *Air Chemistry and Radioactivity,* Academic Press, New York, 1963.

25. R. Jaenicke, "Aerosol physics and chemistry," in *Physical Chemical Properties of the Air, Geophysics and Space Research,* G. Fisher, Ed., Springer-Verlag, Berlin, 1988.

26. G. A. d'Almeida, "On the variability of desert aerosol radiative characteristics," *J. Geophys. Res.,* vol. 93, pp. 3017–3026, 1987.

27. E. P. Shettle, "Optical and radiative properties of a desert aerosol model," in *Proc. Symposium on Radiation in the Atmosphere,* G. Fiocco, Ed., A. Deepak Publishing, Hampton, VA, pp. 74–77, 1984.

28. L. A. Remer and Y. J. Kaufman, "Dynamic aerosol model: Urban/industrial aerosol," *J. Geophys. Res.,* vol. 103, pp. 13,859–13,871, 1998.

29. P. J. Crutzen and M. O. Andreae, "Biomass burning in the tropics: Impact on atmospheric chemistry and biogeochemical cycles," *Science,* vol. 250, pp. 1669–1678, 1990.

30. J. M. Rosen and D. J. Hofmann, "Optical modeling of stratospheric aerosols: Present status," *Appl. Opt.,* vol. 25, no. 3, pp. 410–419, 1986.

31. K. Y. Whitby, "The physical characteristics of sulfur aerosols," *Atmos. Environ.,* vol. 12, pp. 135–159, 1978.

32. S. S. Butcher and R. J. Charlson, *Introduction to Air Chemistry,* Academic Press, New York, 1972.

33. R. D. Cadle, *Particles in the Atmosphere and Space,* Van Nostrand Reinhold, New York, 1966.

34. E. P. Shettle and R. W. Fenn, "Models for the aerosols of the lower atmosphere and the effects of humidity variations on their optical properties," AFGL-TR-79-0214, 1979.

35. B. Herman, A. J. LaRocca, and R. E. Turner, "Atmospheric scattering," in *Infrared Handbook*, W. L. Wolfe and G. J. Zissis, Eds., Environmental Research Institute of Michigan, 1989.

36. V. E. Zuev and G. M. Krekov, *Optical Models of the Atmosphere*, Gidrometeoizdat, Leningrad, 1986.

37. P. V. Hobbs, D. A. Bowdle, and L. F. Radke, "Particles in the lower troposphere over the high plains of the United States. I: Size distributions, elemental compositions and morphologies," *J. Climate Appl. Meteorol.*, vol. 24, pp. 1344–1349, 1985.

38. G. S. Kent, P.-H. Wang, M. P. McCormick, and K. M. Skeens, "Multiyear stratospheric aerosol and gas experiment II measurements of upper tropospheric aerosol characteristics," *J. Geophys. Res.*, vol. 98, pp. 20,725–20,735, 1995.

39. A. Berk, L. S. Bernstein, and D. C. Robertson, "MODTRAN: A moderate resolution model for LOWTRAN 7," Air Force Geophysics Laboratory Technical Report GL TR-89-0122, Hanscom AFB, Bedford, MA, 1989.

40. A. S. Jursa, Ed., *Handbook of Geophysics and the Space Environment*, Air Force Geophysics Laboratory, 1985.

41. C. E. Junge, "Atmospheric chemistry," in *Advances in Geophysics*, vol. 4, Academic Press, New York, 1958.

42. R. A. McClatchey, R. W. Fenn, J. E. A. Selby, F. E. Volz, and J. S. Garing, *Optical Properties of the Atmosphere*, Air Force Cambridge Research Lab, Hanscom AFB, Bedford, MA, AFCRL-72-0497, 1972.

43. D. Deirmenjian, *Electromagnetic Scattering on Spherical Polydispersions*, American Elsevier Publishing, New York, 1969.

44. L. F. Richardson, *Weather Prediction by Numerical Process*, Cambridge University Press, Cambridge, UK, 1922.

45. A. N. Kolmogorov, "The local structure of turbulence in incompressible viscous fluids for very large Reynolds numbers," in *Turbulence, Classic Papers on Statistical Theory*, S. K. Friedlander and L. Topper, Eds., Wiley-Interscience, New York, pp. 151–155, 1961.

46. V. I. Tatarskii, *The Effects of the Turbulent Atmosphere on Wave Propagation*, Translated for NOAA by the Israel Program for Scientific Translations, Jerusalem, 1971.

47. R. H. Kraichman, "On Kolmogorov's inertial-range theories," *J. Fluid Mech.*, vol. 62, pp. 305–330, 1974.

48. A. M. Obukhov, "Temperature field structure in a turbulent flow," *Izv. Acad. Nauk SSSR Ser. Geog. Geofiz.*, vol. 13, pp. 58–69, 1949.

49. R. J. Hill and S. F. Clifford, "Modified spectrum of atmospheric temperature fluctuations and its application to optical propagation," *J. Opt. Soc. Am.*, vol. 68, pp. 892–899, 1978.

50. R. J. Hill, "Models of the scalar spectrum for turbulent advection," *J. Fluid Mech.*, vol. 88, pp. 541–662, 1978.

51. F. H. Champagne, C. A. Friehe, J. C. LaRue, and J. C. Wyngaard, "Flux measurements, flux-estimation techniques, and fine-scale turbulence measurements in the unstable surface layer over land," *J. Atmos. Sci.*, vol. 34, pp. 515–530, 1977.

52. R. M. Williams, Jr., and C. A. Paulson, "Microscale temperature and velocity spectra in the atmospheric boundary layer," *J. Fluid Mech.*, vol. 83, pp. 547–567, 1977.

53. L. C. Andrews, "An analytical model for the refractive index power spectrum and its application to optical scintillations in the atmosphere," *J. Mod. Opt.*, vol. 39, pp. 1849–1853, 1992.

54. J. C. Kaimal, J. C. Wyngaard, D. A. Haugen, O. R. Cote, and Y. Izumi, "Turbulence structure in the convective boundary layer," *J. Atmos. Sci.*, vol. 33, pp. 2152–2161, 1976.

55. V. P. Kukharets and L. R. Tsvang, "Structure parameter of the refractive index in the atmospheric boundary layer," *Izv. Atmos. Oceanic Phys.*, vol. 16, pp. 73–80, 1980.

56. M. G. Miller and P. L. Zieske, "Turbulence environment characterization," RADC-79-131, ADA 072379, Rome Air Development Center, 1979.

57. R. R. Beland, J. H. Brown, R. E. Good, and E. A. Murphy, "Optical turbulence characterization of AMOS, 1985," AFGL-TR-88-0153, Air Force Geophysics Lab, 1988.

58. R. R. Beland, "Propagation through Atmospheric Optical Turbulence," in *The Infrared and Electro-Optical Systems Handbook*, F. G. Smith, Ed., vol. 2, SPIE Press, Bellingham, WA, p. 157, 1993.

59. P. B. Ulrich, "Hufnagel-Valley profiles for specified values of the coherence length and isoplanatic angle," MA-TN-88-013, W. J. Schafer Assoc., 1988.

60. T. E. Van Zandt, G. L. Green, K. S. Gage, and W. L. Clark, "Vertical profiles of refractivity turbulence structure constant: Comparison of observations by the Sunset Radar with a new theoretical model," *Radio Sci.*, vol. 13, pp. 819–827, 1978.

61. J. M. Warnock and T. E. Van Zandt, "A statistical model to estimate the refractivity turbulence structure constant C_n^2 in the free atmosphere," NOAA Tech. Memo, ERL AL-10, Environmental Res. Lab., Boulder, CO, 1985.

62. R. R. Beland and J. H. Brown, "A deterministic temperature model for stratospheric optical turbulence," *Physica Scripta*, vol. 37, pp. 419, 1988.

63. E. M. Dewan, R. E. Good, B. Beland, and J. Brown, *A model for C_n^2 (optical turbulence) profiles using radiosonde data*, PL-TR-93-2043, Environmental Research Papers, no. 1121, Phillips Lab, Hanscom AFB, Bedford, MA, 1993.

64. A. Zilberman and N. S. Kopeika, "Middle East model of vertical turbulence profile," in *Atmospheric Propagation* II, *Proc. SPIE* 5793, pp. 89–97, 2005.

65. J. L. Bufton, P. O. Minott, and M. W. Fitzmaurice, "Measurements of turbulence profiles in the troposphere," *J. Opt. Soc. Am.*, vol. 62, pp. 1068–1075, 1972.

66. J. L. Bufton, "Comparison of vertical profile turbulence structure with stellar observations," *Appl. Opt.*, vol. 12, pp. 1785–1792, 1973.
67. R. E. Good, B. J. Watkins, A. F. Quesada, J. H. Brown, and G. B. Loriot, "Radar and optical measurements of C_n^2," *Appl. Opt.*, vol. 21, pp. 3373–3379, 1982.
68. H. Luce, M. Crochet, F. Dalaudier, and C. Sidi, "An improved interpretation VHF oblique radar echoes by a direct balloon C_n^2 estimation using a horizontal pair of sensors," *Radio Sci.*, vol. 32, pp. 1261–1270, 1997.
69. V. A. Kluckers, N. J. Woodler, M. A. Adcock, T. W. Nicholls, and J. C. Dainty, "Results from SCIDAR experiments," in *Image Propagation through the Atmosphere*, C. Dainty and L. R. Bissonnette, Eds., *Proc. SPIE*, vol. 2828, pp. 234–242, 1996.
70. J. Vernin, M. Crochet, M. Azouit, and O. Ghebrebrhan, "SCIDAR radar simultaneous measurements of atmospheric turbulence," *Radio Sci.*, vol. 25, pp. 953–962, 1990.
71. C. E. Coulman, J. Vernin, and A. Fuchs, "Optical seeing — mechanism of formation of thin turbulent laminae in the atmosphere," *Appl. Opt.*, vol. 34, pp. 5461–5468, 1995.
72. J. C. Ricklin, S. M. Hammel, F. D. Eaton, and S. L. Lachinova, "Atmospheric channel effects on free-space laser communication," *J. Opt. Fiber. Commun.*, vol. 3, pp. 111–158, 2006.
73. J. W. Hardy, *Adaptive Optics for Astronomical Telescopes*, Oxford University Press, 1998.
74. C. E. Coulman, J. Vernin, Y. Coqueugniot, and J. L. Caccia, "Outer scale of turbulence appropriate to modeling refractive-index structure profiles," *Appl. Opt.*, vol. 27, pp. 155–160, 1988.
75. A. Ziad, M. Schock, G. A. Chanan, M. Troy, R. Dekany, B. F. Lane, J. Borgnino, and F. Martin, "Comparison of measurements of the outer scale of turbulence by three different techniques," *Appl. Opt.*, vol. 43, pp. 2316–2324, 2004.
76. A. Muschinski, "Turbulence and gravity waves in the vicinity of a midtropospheric warm front: a case study using VHF echo-intensity measurements and radiosonde data," *Radio Sci.*, vol. 32, pp. 1161–1178, 1997.
77. W. W. Brown, M. C. Roggemann, T. J. Schulz, T. C. Havens, J. T. Beyer, and L. J. Otten, "Measurement and data-processing approach for estimating the spatial statistics of turbulence-induced index of refraction fluctuations in the upper atmosphere," *Appl. Opt.*, vol. 40, pp. 1863–1871, 2001.
78. A. Agabi, J. Borgnino, F. Martin, A. Tokbvinin, and A. Ziad, "G.S.M.: A grating scale monitor for atmospheric turbulence measurements. II First measurements of the wave front outer scale at O.C.A.," *Astron. Astrophys. Suppl. Ser.*, vol. 100, pp. 557–562, 1995.
79. A. Ziad, R. Conan, A. Tokovinin, F. Martin, and J. Borgnino, "From the grating scale monitor to the generalized seeing monitor," *Appl. Opt.*, vol. 39, pp. 5415–5425, 2000.

80. F. Martin, A. Tokovinin, A. Ziad, R. Conan, J. Borgnino, R. Avila, A. Agabi, and M. Sarazin, "First statistical data on wavefront outer scale at La Silla observatory from the GSM instrument," *Astron. Astrophys.*, vol. 336, pp. L49–L52, 1998.

81. R. Conan, R. Avila, L. J. Sanchez, A. Ziad, F. Martin, J. Borgnino, O. Harris, S. I. Gonzalez, R. Michel, and D. Hiriart, "Wavefront outer scale and seeing measurements at San Pedro Martir Observatory," *Astron. Astrophys.*, vol. 396, pp. 723–730, 2002.

82. J. R. Kerr, "Experiments on turbulence characteristics and multiwavelength scintillation phenomena," *J. Opt. Soc. Am.*, vol. 62, pp. 1040–1049, 1972.

83. R. S. Lawrence, G. R. Ochs, and S. F. Clifford, "Measurements of atmospheric turbulence relevant to optical propagation," *J. Opt. Soc. Am.*, vol. 60, pp. 826–830, 1970.

84. E. Vilar and J. Haddon, "Measurement and modeling of scintillation intensity to estimate turbulence parameters in an Earth–space path," *IEEE Trans. Antennas Propag.*, vol. AP-32, pp. 340–346, 1984.

85. A. S. Gurvich and M. S. Belen'kii, "Influence of stratospheric turbulence on infrared imaging," *J. Opt. Soc. Am.*, vol. A 12, pp. 2517–2522, 1995.

86. E. Golbraikh, H. Branover, N. S. Kopeika, and A. Zilberman, "Non-Kolmogorov atmospheric turbulence and optical signal propagation," *Nonlin. Processes Geophys.*, vol. 13, pp. 297–301, 2006.

87. F. Dalaudier, M. Crochet, and C. Sidi, "Direct comparison between in situ and radar measurements of temperature fluctuation spectra: A puzzling result," *Radio Sci.*, vol. 24, pp. 311–324, 1989.

88. F. Daludier and C. Sidi, "Direct evidence of sheets in the atmospheric temperature field," *J. Atmos. Sci.*, vol. 51, pp. 237–248, 1994.

89. H. Luce, F. Daludier, M. Crochet, and C. Sidi, "Direct comparison between in situ and VHF oblique radar measurements of refractive index spectra: A new successful attempt," *Radio Sci.*, vol. 31, pp. 1487–1500, 1996.

90. M. S. Belen'kii, S. J. Karis, J. M. Brown, and R. Q. Fugate, "Experimental study of the effect of non-Kolmogorov stratospheric turbulence on star image motion," *Proc. SPIE*, vol. 3126, pp. 113–123, 1997.

91. M. S. Belen'kii, J. D. Barchers, S. J. Karis, C. L. Osmon, J. M. Brown, and R. Q. Fugate, "Preliminary experimental evidence of anisotropy of turbulence and the effect of non-Kolmogorov turbulence on wavefront tilt statistics," *Proc. SPIE*, vol. 3762, pp. 396–406, 1999.

92. A. Zilberman, E. Golbraikh, N. S. Kopeika, A. Virtser, I. Kupershmidt, and Y. Shtemler, "Lidar study of aerosol turbulence characteristics in the troposphere: Kolmogorov and non-Kolmogorov turbulence," *Atmos. Res.*, vol. 88, pp. 66–77, 2008.

93. E. Golbraikh and N. Kopeika, "Behavior of structure function of refraction coefficients in different turbulent fields," *Appl. Opt.*, vol. 43, pp. 6151–6156, 2004.

94. G. K. Batchelor, "Small-scale variation of convected quantities like temperature in turbulent fluid. Part I. General discussion and the case of small conductivity," *J. Fluid Mech.*, vol. 5, pp. 113–133, 1959.

95. C. H. Gibson and W. H. Schwarz, "The universal equilibrium spectra of turbulent velocity and scalar fields," *J. Fluid Mech.*, vol. 16, pp. 365–384, 1963.

96. C. H. Gibson, W. T. Ashurst, and A. R. Kerstein, "Mixing of strongly diffusive passive scalars like temperature by turbulence," *J. Fluid Mech.*, vol. 194, pp. 261–293, 1988.

97. A. Brissaud, U. Frisch, J. Leorat, M. Lesieur, and A. Mazure, "Helicity cascades in fully developed isotropic turbulence," *Phys. Fluids*, vol. 16, pp. 1366–1367, 1973.

98. S. S. Moiseev and O. G. Chkhetiani, "Helical scaling in turbulence," *JETP*, vol. 83, pp. 192–198, 1996.

99. T. Elperin, N. Kleeorin, and I. Rogachevskii, "Isotropic and anisotropic spectra of passive scalar fluctuations in turbulent fluid flow," *Phys. Rev.*, vol. E 53, pp. 3431–3441, 1996.

100. A. Zilberman, E. Golbraikh, and N. S. Kopeika, "Lidar studies of aerosols and non-Kolmogorov turbulence in the Mediterranean troposphere," *Proc. SPIE*, vol. 5987, pp. 15–26, 2005.

101. E. Golbraikh and N. Kopeika, "Turbulence strength parameter in laboratory and natural optical experiments in non-Kolmogorov cases," *Opt. Commun.*, vol. 242, pp. 333–338, 2004.

102. E. Golbraikh and N. Kopeika, "Changes in modulation transfer function and optical resolution in helical turbulent media," *J. Opt. Soc. Am.*, vol. A 19, pp. 1774–1778, 2002.

103. B. E. Stribling, B. M. Welsh, and M. C. Roggemann, "Optical propagation in non-Kolmogorov atmospheric turbulence," *Proc. SPIE*, vol. 2471, pp. 181–196, 1995.

2

OPTICAL WAVE PROPAGATION IN THE ATMOSPHERE

Let us consider the effects of various structures of the layered gaseous inhomogeneous atmosphere on optical signal propagation and the corresponding processes that accompany it, such as refraction, absorption, attenuation, and scattering [1–29].

2.1 Refraction Phenomena

Refraction occurs as a result of propagation effects through the layered quasi-homogeneous structure of the atmosphere, as a gaseous continuum, causing optical rays to propagate not along the straight radiopaths but to curve slightly toward the ground. This phenomenon will be described here, considering the atmosphere as a quasi-homogeneous gaseous medium consisting of molecules and atoms of gases.

The propagation properties of the quasi-homogeneous layered atmosphere are characterized by the refraction index, n, related to the dielectric permittivity of the air, ε_r, as $n = \sqrt{\varepsilon_r}$. The refractive index, n, of the Earth's atmosphere is slightly greater than 1, with a typical value at the Earth's surface of around 1.0003. Because this value is so close to unity, it is common to express the refractive index in N units, usually called *refractivity* [1, 7, 19], which is the difference between the actual value of the refractive index and unity in parts per million:

$$N = (n - 1) \cdot 10^6 \qquad (2.1)$$

Thus, at the ground surface the refractivity is $N = N_S \approx 315$ N units.

In the real atmosphere, refractivity, N, varies with gas and water vapor pressure and with gas temperature. The variations of temperature, pressure, and humidity from point to point within the troposphere cause the variations of the refractivity, N, which can be estimated according to the semi-empirical Debye formula [1, 7, 19]:

$$N = \frac{77.6}{T}(p_A + 4810 \cdot p_W T^{-1}) \tag{2.2}$$

where T is the absolute temperature in Kelvin (K), p_A is the atmospheric pressure in millibars, and p_W is the water vapor pressure in millibars.

Usually, the troposphere can be considered as a spherically layered quasi-homogeneous medium, where the dominant variations of N are vertical (e.g., along the altitude above ground surface) and the parameter N reduces toward zero (n becomes close to unity) with increasing altitude. These variations are approximately exponential within the first few tens of kilometers of the Earth's atmosphere (i.e., within the troposphere [1]):

$$N = N_s \exp\left\{-\frac{h}{H}\right\} \tag{2.3}$$

where h is the height above sea level and $N_s \approx 315$ is the standard reference value. From Equation 2.3, the parameter $H = 7.35$ km was defined as the height scale of the standard atmosphere. Equation 2.3 is called a standard exponential model of the troposphere, valid with the assumption that the temperature of the air decreases linearly with height and the humidity decreases exponentially.

The refractive index variation with height according to Equation 2.3 causes the phase velocity of optical waves to be slightly slower closer to the Earth's surface, such that the ray paths are not straight but tend to curve slightly toward the ground. In other words, the elevation angle, α_1, of an initial ray at any arbitrary point (see Figure 2.1) is changed after refraction to an angle α_2.

The same situation will exist at the next virtual layer of the atmosphere with a different refractive index n. Finally, the ray launched from the Earth's surface propagates over the curve, and the radius of curvature at any point ρ is given in terms of the derivative of n for height [1, 7, 19]:

$$\rho = -\left(\frac{\cos\alpha_1}{n}\frac{dn}{dh}\right)^{-1} \tag{2.4}$$

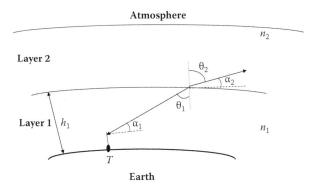

Figure 2.1   Sketched refraction phenomenon caused by layered atmosphere. (From N. Blaunstein and C. Christodoulou, *Radio Propagation and Adaptive Antennas for Wireless Communication Links: Terrestrial, Atmospheric, and Ionospheric*, Wiley InterScience, Hoboken, NJ, 2006. Reprinted with permission of John Wiley & Sons, Inc.)

Usually, the gradient of the refractivity, $g(h) = dN/dh$, can be assumed exponential near the Earth's surface [1, 7]:

$$g_s(h) = -0.04\exp(-0.136\,h),\ \text{km}^{-1} \tag{2.5}$$

However, in the first approximation we can, following the work in References 1, 7, and 19, use the linear model, assuming the gradient to be a constant equal to its value at $h = 0$: $g = g(0)$. Then, the standard atmosphere of Equation 2.3 can be approximated as linear:

$$N \approx N_S - \frac{N_S}{H}h \tag{2.6}$$

As was found experimentally (see References 1 and 7), the refractivity thus has a nearly constant gradient of about −43 N units per kilometer, starting at the ground surface (i.e., from $h = 0$) from the standard reference value of refractivity of $N_S = 315$ N units.

2.2 Effects of Aerosols

Aerosols affect mostly absorption and scattering of optical waves passing through the inhomogeneous atmosphere. Absorption occurs as a result of conversion from optical wave energy to thermal energy within an attenuating particle.

2.2.1 Attenuation of Aerosols

We begin by considering the optical wave attenuation caused by the atmospheric gas as a continuum consisting of molecules of gases and aerosols. The absorption in the atmosphere over a path length r is given by [2, 4, 7]

$$A = \int_0^r dr\, \gamma(r), \text{ dB} \tag{2.7}$$

where $\gamma(r)$ is the specific attenuation consisting of two components:

$$\gamma(r) = \gamma_o(r) + \gamma_w(r), \text{ dB/km} \tag{2.8}$$

where $\gamma_o(r)$ and $\gamma_w(r)$ are the contributions of oxygen and water vapor, respectively.

Because in meteorology the measurable quantity is the relative humidity, $\eta(T)$, we have to relate ρ with $\eta(T)$. The relative humidity is given by

$$\eta(T) = p/E(T) \tag{2.9}$$

where p is the water vapor partial pressure (in millibars) and $E(T)$ is the saturation pressure, which is defined by the approximate formula [2–4, 7]

$$E(T) \cong 24.1\Theta^5\, 10^{10-9.834\Theta} \tag{2.10}$$

where $\Theta = 300/T$. Combining the above formulas, in Reference 7 it was found that the relationship between the values of p and ρ is

$$\rho \cong 216.7\, p/T \tag{2.11}$$

The total atmospheric attenuation, L_a, for a particular path is then found by integrating the total specific attenuation over the total path length, r_T, in the atmosphere [2–4, 7]:

$$L_a = \int_0^{r_T} \gamma_a(l)dl = \int_0^{r_T} [\gamma_w(l) + \gamma_o(l)]dl, \text{ dB} \tag{2.12}$$

2.2.2 Scattering by Aerosols

The attenuation due to scattering of the optical wave depends on the field of view (FOV) of the optical receiver. If the FOV is very large, some field energy scattered at very small forward angles will still be received and detected. If the FOV is very small, virtually all scattered radiation can be rejected and only transmitted radiation arrives at the detector.

For optical waves passing through the inhomogeneous atmosphere, two approaches can be used according to the relationship between the wavelength of the relation being scattered and the size of the particles causing the scatter. These approaches are (1) Mie scattering and (2) nonselective scattering [8–11].

Mie scattering is applicable where the particle size is *comparable* to the radiation wavelength. The Mie scattering area coefficient is defined as the ratio of the incident wavefront that is affected by the particle to the cross-sectional area of the particle itself. The scattering coefficient, σ, can be obtained from References 8–11:

$$\sigma = NK\pi a^2 \tag{2.13}$$

where the value of K rises from 0 to nearly 4 and asymptotically approaches the value 2 for large droplets. For the almost universal condition in which there is a continuous size distribution in the particles, we have from Reference 8:

$$\sigma_\lambda = \pi \int_{a_1}^{a_2} N(a)K(a,n)a^2 da \tag{2.14}$$

where σ_λ is the scattering coefficient for wavelength (cm^{-1}), $N(a)$ is a number of particles per cubic centimeter in the interval da, (cm^{-3}), $K(a,n)$ is the scattering area coefficient, a is the radius of spherical particles, and n is an index of refraction of particles. To convert the results of the computations of Equation 2.14 to km^{-1}, the more commonly used unit for σ is that which is multiplied by 10^5. Many authors have presented a detailed treatment of scattering theory. Thus, in References 8–11, application of Mie theory for a wide variety of particle composition, size, and shape is discussed.

Nonselective scattering occurs when the *particle size is very much larger* than the radiation wavelength. Large-particle scattering is composed of contributions from three processes involved in the interaction of the optical radiation with the scattering particles:

1. Reflection from the surface of the particle with no penetration
2. Passage through the particle with and without internal reflections
3. Diffraction at the edge of the particle

In References 8–11, the combined effect of all three processes, including the interference encountered among the three components, is discussed. It is shown that for particles larger than about two times the wavelength of the radiation ($a > 20$), the scattering area coefficient becomes equal roughly to 2, which is the asymptotic value approached by the Mie coefficient. Thus, the theoretical approach through diffraction, refraction, and reflection appears to have little contribution to the more general approach of Mie. For $a < 20$, the Mie theory is valid, and for $a > 20$, the two predictions converge on the value 2.

2.3 Aerosol Effects on Optical Wave Propagation

Several different scattering processes occur in the atmosphere as a result of the mixture of sizes of particles, which range from molecules to aerosols, ice crystals, and water droplets.

The ratio of particle size to wavelength of light is one of the parameters determining the light scattering characteristics. The theory employs the size parameter $a = 2\pi r/\lambda$, where r is the particle radius and λ is the wavelength. This expression assumes a spherical shape of the particle.

If the particle is large compared with the wavelength of the incident light, the scattering process can be described by geometric optics laws. Those particles that are very small compared to the wavelength will produce Rayleigh scattering. The scatter process in Rayleigh scattering is proportional to λ^{-4}. In between, selective scattering occurs (λ^{-x}, $0 < x < 4$) and the Mie theory is used as the method of analysis [12, 13].

The Mie theory treat the scattering of light by a homogeneous spherical particle of arbitrary size and refractive index. The Mie solution for scatter from a spherical particle results in infinite series expressions for the field components polarized perpendicular and parallel to the scattering plane. The solution for the spatial distribution of scattered light for incident unpolarized natural light is of the form [8–10, 30, 31]

$$I_\lambda = I_{0,\lambda} \frac{1}{k^2 R^2} \cdot \frac{i_1(a,m,\theta) + i_2(a,m,\theta)}{2}, \qquad (2.15)$$

where I_λ is the intensity of scattered light (power per unit area); $I_{0,\lambda}$ is the intensity of incident light; i_1 and i_2 are the intensity functions; $k = 2\pi/\lambda$, where λ is the wavelength of incident light; R is the distance from the scattering particle center; $a = 2\pi r/\lambda$ is the size parameter, where r is the particle radius; m is the refractive index of the particle; and θ is the angle between the incident and the scattered beams ($\theta = 0^0$ is defined as forward scattering).

The dimensionless intensity functions i_1 and i_2 are proportional to the square of the electric-field components perpendicular and parallel, respectively, to the plane of observation. Experimentally they can be observed separately by inserting a polarizer in the scattered beam.

The scattering properties of aerosol particles depend on the refractive index, m, of the particle material, which in general form has real and imaginary parts; i.e., $m = n_r - jn_i$. The absorption properties of a particle are determined by the imaginary part of the refractive index, n_i. Small changes in the imaginary part of the refractive index can have large effects on the total extinction. For nonabsorbing materials, the refractive index is a real number.

The total energy scattered by particle may be calculated by the usual integration over the sphere:

$$F_s = \int_0^{4\pi} I d\Omega = I_0 C_s \qquad (2.16)$$

This expression gives the lumens scattered by one particle per lumen/cm² of illuminance (or watts per particle per watt/cm² irradiance). That is, the total scattered light, F_s, is equal to the total light, I_0, incident on C_s, where C_s is the "effective area" or scattering cross-section of the particle (the area of wavefront that is affected by particle).

The scattering cross-section is represented by intensity functions [30, 32]

$$C_S = \frac{1}{I_0} \int I \, d\Omega = \frac{\lambda^2}{2\pi} \int_0^\pi [i_1(\theta, m, a) + i_2(\theta, m, a)] \sin\theta \, d\theta. \quad (2.17)$$

where $d\Omega = 2\pi \sin\theta d\theta$. The scattering, absorption, and extinction cross-sections are related as $C_s + C_a = C_{ext}$. Thus, without absorption the extinction cross-section is equal to the scattering cross-section.

The ratio of the scattering cross-section to the geometrical cross-section of particle πr^2 is called the scattering area ratio or efficiency factor for scattering, $Q_s = C_s/\pi r^2$. Efficiency factors Q_a and Q_{ext} are defined similarly. Q_{ext} (or Q_s) is a dimensionless quantity that actually depends on r and λ only through the ratio $a = 2\pi r/\lambda$. For small a, when m is real, Q_{ext} increases as a^4, which corresponds to Rayleigh scattering. Then it reaches a maximum for r that is comparable with λ and, for large a, tends oscillatorily toward a value 2, which corresponds to the diffraction limit.

The real part of the refraction index of the particle affects the positions of the principal and secondary peaks of Q_{ext}. The absorption properties of a particle are determined by the imaginary part of the refractive index, n_i. Small changes in the imaginary part of the refractive index smear out the Q_{ext} structure and depress the amplitude of its oscillations. The scattering efficiency maximum for different λ is located at different radii values. It is fundamental in the use of spectral extinction data for particle sizing. The scattering process goes thus [33]:

$$\lambda^{-4} \text{ for Rayleigh } (r \ll \lambda)$$

$$\lambda^{-2} \text{ for Mie } (r \sim \lambda/2)$$

$$\lambda^{-1} \text{ for Mie } (r \sim 3\lambda/4)$$

$$\lambda^0 \text{ for Mie } (r \sim \lambda)$$

Nonselective scattering occurs when $r \gg \lambda$ and is the combination of reflection, refraction, and diffraction. The scattering coefficient for monodisperse aerosol, which is equal to the extinction coefficient if no absorption is assumed, is given by the relation

$$\sigma_\lambda = N \cdot Q(\lambda, m, r) \cdot \pi r^2, \quad (2.18)$$

where N is the concentration of particles (cm^{-3}).

The relationship between the size distribution of the aerosol particles and their spectral extinction coefficient, σ_λ, can be expressed in the other manner (Equation 2.14) by the following integral equation [32, 33]:

$$\sigma_\lambda = \int \pi r^2 Q_{ext}(r, \lambda, m) N(r) dr. \tag{2.19}$$

where $N(r)$ is the concentration of particles (cm^{-3}) with radii between r and $r + dr$.

It is convenient to describe the relative angular distribution of the radiant intensity scattered by a volume element by the function called phase function:

$$P(\theta) = \frac{4\pi}{k^2 \sigma_S} \cdot \left[\frac{i_1(a, m, \theta) + i_2(a, m, \theta)}{2} \right], \tag{2.20}$$

where σ_S is the scatter cross-section per unit volume. The scattering phase function describes the direction in which light is deflected. As wavelength decreases the scatter becomes more and more directed into small angles in the forward direction. The values of $i_{1,2}$, $P(\theta)$, and $Q(\lambda, m, r)$ can be calculated using the Mie scattering algorithms [31, 34].

2.3.1 Atmospheric Transmission

For light traveling through the scattering medium containing N particles of radius r per unit volume, the loss of light due to scattering per unit path length dz is

$$dI = -I \cdot C_s N\, dz \tag{2.21}$$

Integration gives the well-known Bouguer extinction law (also called Beer-Lambert law) [35, 36]:

$$I = I_0 \exp(-\sigma_s z) \tag{2.22a}$$

where I_0 is the light intensity of the source and $\sigma_s = C_s N$ is the scattering coefficient. The optical thickness for scattering is $\tau = \sigma_s z$, and the atmospheric transmittance is

$$T = I/I_0 = \exp(-\sigma_s z) \tag{2.22b}$$

This is the solution for the case of propagation through a homogeneous medium. For the case in which the scatter properties of the medium vary along the path length, the exponent is the path integral along the propagation path. It is related to the path optical depth or integrated extinction by

$$T(\lambda, z) = \exp\left(-\int_0^z \sigma(\lambda, z')dz' \right), \tag{2.23}$$

where $\sigma(z)$ is the atmospheric unit volume extinction coefficient (km^{-1}) at any intervening height z' between 0 and z. The extinction coefficient includes the loss effects of both scattering and absorption.

In general, the total extinction coefficient σ_t is a sum of contributions from molecular absorption, α_m, molecular scattering, σ_m, aerosol absorption, α_a, and aerosol scattering, σ_a:

$$\sigma_t = \alpha_m + \sigma_m + \alpha_a + \sigma_a \tag{2.24}$$

The molecular absorption effects can be neglected if the wavelength of the transmitter lies in a spectral transmission window of the atmosphere. The most useful transparent spectral ranges are the visible (0.4–0.7 µm), the near-infrared (0.7–1.5 µm), and the windows between 3–5 µm and 9–13 µm. The aerosol absorption and scattering values at different wavelengths can be found in References 37 and 38.

A commonly used parameter is visual range or "visibility" (also called *meteorological range*) [29, 33]. It corresponds to the range at which 550 nm radiation is attenuated to 0.02 times its transmitted level and is equal to $3.912/\sigma_{550}$, where σ_{550} is the atmospheric attenuation coefficient at 550 nm wavelength. Here Rayleigh scattering by molecules implies a visual range of about 340 km. This situation is shown on the left in Figure 2.2.

Thus, Rayleigh scattering theory applies only to extremely clear air. The Rayleigh scattering coefficient for molecular scattering is therefore an extreme limit that can never be exceeded in practice.

2.3.2 Aerosol Beam Widening

As the laser beam propagates through the scattering medium, the photons are multiply scattered within the medium. As a result of this

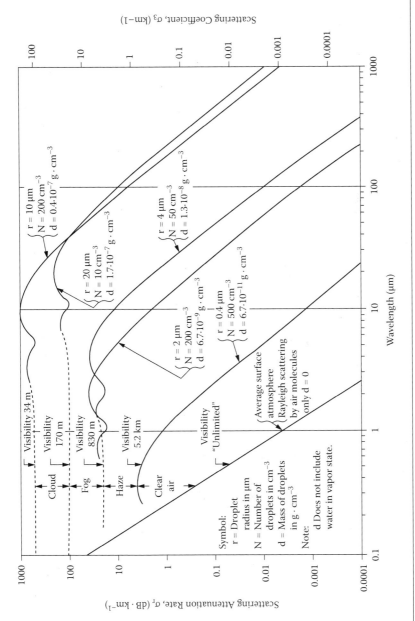

Figure 2.2   Scattering coefficient for a variety of weather conditions.

multiple-scatter effect, the original radiance distribution undergoes angular and spatial spreading that reduces beam irradiance at the target.

The average irradiance, <*I*> (watts/cm²), at a distance *r* from the optical axis of the transmitter in a target plane located at range *z* is given (without turbulence influence) by [39]

$$< I(r,z) > \approx \exp[-\tau] \cdot \frac{P_T}{\pi W_{aer}^2} \cdot \exp\left[-r^2/W_{aer}^2\right], \qquad (2.25)$$

where P_T is the transmitted power, $\tau = \sigma_s z$ is the optical depth, and W_{aer} is the resulting beam widening (effective beam radius) created by forward-biased multiple scattering. It is given (for a collimated Gaussian beam) by

$$W_{aer}(z) = \left(W_s^2(z) + W^2(z)\right)^{1/2}, \qquad (2.26)$$

where *W* is the beam widening due to diffraction effects and W_s is the contribution to the beam width due to forward multiple aerosol scattering. It is given by

$$W^2 = W_0^2\left[1 + \left(\frac{z}{kW_0^2}\right)^2\right], \qquad (2.27)$$

$$W_S^2 = \frac{z^2 \cdot \tau \cdot \langle \theta_0^2 \rangle}{3}, \qquad (2.28)$$

where *z* is the distance to target, $k = 2\pi/\lambda$.

The parameter $\langle \theta_0^2 \rangle$ is the mean square scattering angle associated with the scatter medium under the small angle scattering approximation, equal to $\theta_0^2 = 1/\pi P(0)$, where $P(0)$ is the scalar phase function of the scattering medium evaluated at $\theta_0 = 0$ [39–41]. Here it was assumed that the phase function is approximated as a Gaussian function of scattering angle for forward-scattering events. The key to estimating beam widening by aerosols is the aerosol size distribution. Equation 2.28 is valid in the case of $\tau \gg 1$, which may correspond to high aerosol loading or long propagation distances.

2.4 Effects of Hydrometeors

The main influenced particles in optical links passing through the atmosphere are hydrometeors, including raindrops, fog, snow, clouds and so on. Hydrometeors mostly affect absorption and scattering of optical waves propagating through the inhomogeneous atmosphere. Effects of absorption were considered in a previous section. Now we briefly consider the effects of scattering. Scattering, as a propagation phenomenon, occurs from redirection of the optical waves into various directions, so that only a fraction of the incident energy is transmitted onward in the direction of the detector [2, 19]. This process is frequency dependent, because wavelengths that are long when compared to the particles' size will be only weakly scattered.

2.4.1 Effects of Clouds and Fog

As follows from numerous observations, in clouds and fog the drops are always smaller than 0.1 mm and the theory for the small size scatterers is applicable [18, 19]. This gives

$$\gamma_c \approx 0.438\, c(t)q/\lambda^2, \text{ dB/km} \qquad (2.29)$$

where λ is the wavelength measured in centimeters and q is the water content measured in g/m³. For visibility of 600 m, 120 m, and 30 m, the water content in fog or clouds is 0.032 g/m³, 0.32 g/m³, and 2.3 g/m³, respectively. Calculations show that attenuation in moderately strong fog or clouds does not exceed attenuation due to rain with a rainfall rate of 6 mm/h. Because there are not even statistics at one point of the liquid water content in clouds and fog, and, what is more, no empirical data on the path-averaged values are known, we have to restrict ourselves to a semi-heuristic model only. Specifically, we assume that the thickness of a cloud layer is $w_c = 1$ km and the lower boundary of the layer is located at $h_c = 2$ km height. The water content of clouds has a yearly percentage of the form [19]

$$P(x > L_c) = p_c \exp(-0.56\sqrt{L_c} - 4.8\,L_c), \% \qquad (2.30)$$

where p_c is the probability of cloudy weather (%) and L_c is the length of the beam path within the cloud layer. Neglecting ray bending (we will discuss this phenomenon later), we have for the length of the path within the cloud layer:

$$L_c = 0, \quad h_2 \le h_c \tag{2.31a}$$

$$L_c = \sqrt{d^2 + h_2^2}\,(1 - h_c/h_2), \quad h_c < h_2 < h_c + w_c \tag{2.31b}$$

$$L_c = w_c/\sin\theta, \quad h_2 \ge h_c + w_c \tag{2.31c}$$

where $\theta = \arctan(h_2/d)$. Here, as before, h_2 is the air vehicle (receiver or transmitter) height. Although attenuation in clouds is less than in rain, the occurrence of clouds can be much more essential than that of rain.

2.4.2 Effects of Rain

The attenuation of optical waves caused by rain increases with the number of raindrops along the ray trace, the size of the drops, and the length of the path through the rain.

If such parameters of rain as drop density and size in a given area containing rain are constant, then, according to References 5, 7, and 12, the signal power, P_r, at the receiver decreases exponentially with the length of the ray trace, r, through the rain with the parameter of power attenuation in e^{-1} times, denoted by α, that is,

$$P_r = P_r(0)\exp\{-\alpha r\} \tag{2.32}$$

Expressing Equation 2.18 in logarithmic scale gives path loss:

$$L = 10\log\frac{P_t}{P_r} = 4.343\alpha r \ \text{dB} \tag{2.33}$$

Another way to estimate the total loss via the specific attenuation, in dB/m, was shown in References 7 and 12. We define this factor as

$$\gamma = \frac{L}{r} = 4.343\alpha \ \text{dB/m} \tag{2.34}$$

where now the power attenuation factor, α, can be expressed as the integral effects of the one-dimensional (1D) distribution of drop diameter, D, denoted by $N(D)$, and the effective cross-section of frequency-dependent optical signal power attenuation by raindrops,

$C(D)$ (dB/m); i.e.,

$$\alpha = \int\limits_{D=0}^{\infty} N(D) \cdot C(D) dD \qquad (2.35)$$

As mentioned in References 5, 7, and 13, in real situations in the atmosphere the drop diameter distribution, $N(D)$, is not a constant, and one must consider the range dependence of the specific attenuation [i.e., the range dependence $\gamma = \gamma(r)$] and integrate it over the whole ray trace length, r_R, to find the total path loss:

$$L = \int\limits_{0}^{r_R} \gamma(r) dr \qquad (2.36)$$

To solve Equation 2.36, in Reference 13 a special mathematical procedure that considered a particular drop size distribution was proposed. It gave, before integration of Equation 2.36, the following expression for $N(D)$:

$$N(D) = N_0 \exp\left\{-\frac{D}{D_m}\right\} \qquad (2.37)$$

where N_0 and D_m are parameters, with D_m depending on the rainfall rate, R, measured on the ground in mm/h, with $N_0 = 8 \cdot 10^3$ m^{-2} mm^{-1} and

$$D_m = 0.122 \cdot R^{0.21} \text{ mm} \qquad (2.38)$$

As for the attenuation cross-section, $C(D)$, from Equation 2.35, it can be found using the Rayleigh approximation, which is valid for lower frequencies; that is, for the case when the average drop size is smaller compared to the ray wavelength. According to the Rayleigh approximation [7], we have a very simple expression for $C(D) \propto D^3/\lambda$, where λ is the wavelength.

2.5 Effects of Atmospheric Turbulence on Optical Propagation

As mentioned in Chapter 1, atmospheric turbulence is air motion that represents a set of vertices or the turbulent structures (called *eddies*) of various scale sizes, extending from large-scale size to smaller ones.

In addition, atmospheric turbulence is a product of irregular air movement in which the wind constantly varies in speed and direction. Churns and mixes in the atmosphere cause water vapor, smoke, and other aerosols, as well as energy, to become distributed at all atmospheric elevations. Due to turbulent flows causing eddies of wind in the troposphere, the mainly horizontal layers of equal refractive indices in it become mixed, leading to rapid refractive index variations over small distances, called *small-scale* variations, as well as over short time intervals, called *rapid* refractive index variations.

When a beam of light passes through the atmosphere the refractive index inhomogeneities affect the beam in at least three different ways. Because of random fluctuations in phase velocity the initially defined phase front becomes distorted. This alters and redirects the flow of energy in the beam and, as a result, random changes in beam direction (beam wander) and intensity fluctuations (scintillation) occur. The beam also spreads in size beyond the dimensions predicted by diffraction theory. In general, beam wander effects arise when the size of turbulence eddies (cells) are larger than the width of the beam that causes the wavefront tilt, whereas those eddies that are small compared with the beam diameter tend to widen the beam.

The task of relating the statistical properties of the atmospheric refractive index to the behavior of the amplitude and phase of optical waves has been attempted in a large number of experimental and theoretical works, and several books and reviews on the subject have appeared in the last several decades [20–26, 42–46].

The most important statistical quantities associated with laser communication systems are mean irradiance and scintillation index. The mean irradiance is used to imply beam spreading beyond pure diffraction, which is important for calculating power losses at the receiver caused by atmospheric turbulence effects. For laser communication systems, the implied power loss through beam spreading will lead to lower signal-to-noise ratios (SNR), which in turn will increase the probability of signal fade and bit error rate (BER) of the system. The scintillation index describes the irradiance fluctuations at the receiver, which is important in calculating signal fades. The scintillation index is considered the most deleterious effect on system performance in laser communication systems. Scintillation reduces the mean SNR in direct detection systems and also increases the probability (and duration) of signal fade and BER.

Over the years, different models have been presented to predict the atmospheric turbulence parameter C_n [19–29, 47–67]. Direct connection between physical parameters of atmospheric communication channels and atmospheric turbulence parameters is given by the Kolmogorov model [45, 47, 66].

In this section, we will discuss briefly the scintillations of propagating waves, as well as the effects of beam wander and angle-of-arrival variations at the receiver due to irregular structure of the atmosphere.

2.5.1 Scintillations

As mentioned, a ray propagating through a turbulent atmospheric medium will experience irradiance fluctuations, called *scintillations* [19–26]. Waves traveling through layers with rapid variations of refractive index therefore vary rapidly and randomly in amplitude and phase. In particular, the important scale sizes in irradiance fluctuations are small scales on the order of the Fresnel zone $[L/k]^{\frac{1}{2}}$ or inner scale l_0, whereas it is the larger scale sizes (including outer scale L_0) that affect phase fluctuations. Scintillations are not an absorptive effect and lead to signal amplitude and phase fluctuations in both the space and time domains.

To establish highly reliable free-space optical (laser) communication links, quantitative estimates of various statistical quantities that are associated with atmospheric turbulence-induced scintillations are necessary. The scintillation index (normalized variance of the intensity fluctuations) σ_I^2 describes fluctuations in optical power as measured by a point receiver. It is related to the log-amplitude variance (Rytov variance), σ_X^2, through [23, 24]

$$\sigma_I^2 = \frac{\langle I^2(\rho, L)\rangle}{\langle I(\rho, L)\rangle^2} - 1 = \exp\left[4\sigma_X^2\right] - 1 \qquad (2.39a)$$

where $I(\rho, L)$ is the intensity received at a point on the receiver aperture after propagating a distance L through the turbulent channel, and angle brackets < > denote ensemble average, which is also equal to the long-time average, assuming the process to be ergodic. In the weak fluctuation regime, where $\sigma_X^2 < 0.3$, it can be approximated by

$$\sigma_I^2 = \exp[4\sigma_X^2] - 1 \approx 4\sigma_X^2 \qquad (2.39b)$$

The variance in log-irradiance, which is the usual parameter that is measured experimentally, can be related to the theoretically derived log-amplitude variance (Rytov variance) by $\sigma_{\ln I}^2 = 4\sigma_X^2$.

Knowledge of σ_I^2 allows a determination of the dynamic range desired by the detector/sensor used to receive the signal in an optical link. The correlation of intensity fluctuations, in a plane perpendicular to the beam, falls to zero for points separated by a distance slightly greater than $[\lambda L]^{\frac{1}{2}}$ and, for a receiver diameter greater than this value, the decreasing correlation across the whole aperture tends to smooth the variations (aperture-averaging effect). Aperture averaging reduces the probability of fades. In Reference 45, an engineering approximation for the magnitude of aperture averaging effects on ground-based receivers is suggested.

As mentioned previously, the strength of scintillation can be defined in terms of the variance of the optical wave log-amplitude or its irradiance at a point. The log-amplitude variance, σ_X^2, can be used to characterize the strength of turbulence for an optical link.

The technique of describing the fluctuations of a propagating wave as log functions (log-amplitude or log-irradiance) arises conveniently from the Rytov method and is described in detail elsewhere (see, for example, References 23, 24, 29, and 42–44). Basic functions for the statistical representation of wave fluctuations are the covariance functions of the amplitude and phase for a plane or spherical wave propagating through a turbulent medium.

The covariance function for the fluctuations of the log-amplitude, X, of an electromagnetic wave is defined by

$$C_X(L, \boldsymbol{\rho}_1, \boldsymbol{\rho}_2) = \langle X(L, \boldsymbol{\rho}_1) \cdot X(L, \boldsymbol{\rho}_2) \rangle \qquad (2.40)$$

where $X = \ln A/A_0$, where A is the amplitude of a component of the electric field, A_0 is its unperturbed value, $\boldsymbol{\rho}_1$ and $\boldsymbol{\rho}_2$ are the two points in a plane transverse to the direction of propagation at range L, and angular brackets denote a time or ensemble average.

By assuming a theoretical refractive index spectrum (Equation 1.45), a relation between the turbulent scales and covariance or variance of the log-amplitude, $C_X(L, 0) \equiv \sigma_X^2$, can be obtained by solving the wave equation. With the Rytov approximation, the covariance due to isotropic and homogeneous refractive index field can be expressed as an integral equation [23]:

$$C_X(L,\rho) = 4\pi^2 k^2 A(\alpha) \int_0^L \beta(z)\,dz \int_0^\infty dK K^{1-\alpha} J_0(bK\rho) \sin^2\left[\frac{bK^2(L-z)}{2k}\right]$$

(2.41)

where $k = 2\pi/\lambda$, $b = 1$ for plane waves, and $b = z/L$ for spherical waves. In Equation 2.41, K is the spatial wave number, z is the coordinate along the propagation path L, $A(\alpha)$ is given by Equation 1.46 (see Chapter 1), α is the power law exponent, and J_0 is the zero-order Bessel function of the first kind. $C_X(L, \rho)$ describes the covariance of the log-amplitude X of the electromagnetic field at two points, separated by ρ in a plane transverse to the propagation path at distance L. Equation 2.41 can be used for theoretical solutions that give measurable quantities for specific applications. $C_X(L, \rho)$ has a correlation distance of the order $[z\lambda]^{\frac{1}{2}}$, the radius of the first Fresnel zone. Turbulence scales of dimensions $[z\lambda]^{\frac{1}{2}}$ located at a distance z from the receiving plane will have a large effect on $C_X(L, \rho)$.

For a fixed path length z, eddies larger than the Fresnel zone (i.e., wavenumbers smaller than about $[z\lambda]^{-\frac{1}{2}}$) contribute primarily to phase and not amplitude scintillations (refractive effect). A theoretical expression for log-amplitude variance, σ_X^2, has been derived by Tatarskii [23] on the basis of the Rytov approximation, the O-K (Obukhov-Kolmogorov) turbulence spectrum (Chapter 1), and with assumption of plane-wave propagation. It has the form [23]

$$\sigma_X^2 = 0.56(\sec\varphi)^{\frac{11}{6}} k^{\frac{7}{6}} \int_0^L C_n^2(z)(z')^{\frac{5}{6}} dz, \quad \text{at } L_0 \gg (\lambda L)^{1/2} \gg l_0 \quad (2.42)$$

where L is the path length, $z' = z$ is the distance along the propagation path as measured from the receiver (downlink), $z' = L - z$ for the uplink, $\varphi \leq 1$ rad is the zenith angle, and $k = 2\pi/\lambda$, with λ being the wavelength of the radiation. L is the height of the receiver (uplink) or transmitter (downlink). For larger angles, φ, this result may not be useful as scintillation effects move into the strong turbulence regime.

The Rytov approximation is limited to regimes where the turbulence is weak and/or the propagation distance is short. For O-K turbulence the Rytov solution has been shown to be valid when the log-amplitude variance is small; i.e., $\sigma_X^2 \leq 0.3$–0.5 [23]. On the other hand, for vertical propagation no such saturation effects have been observed and it is felt that the theory holds well.

By solving Equation 2.41, the generalized form of log-amplitude variance [when $C_X(L, 0) \equiv \sigma_X^2$] can be represented as [46]

$$\sigma_X^2 = B(\alpha)(\sec\varphi)^{\frac{\alpha}{2}} A(\alpha) k^{\frac{6-\alpha}{2}} \int_0^L \beta(z)(L-z)^{\frac{\alpha-2}{2}} dz, \quad 3 < \alpha < 5 \quad (2.43a)$$

for a plane wave, and

$$\sigma_X^2 = B(\alpha)(\sec\varphi)^{\frac{\alpha}{2}} A(\alpha) k^{\frac{6-\alpha}{2}} \int_0^L \beta(z)\left(\frac{z}{L}\right)^{\frac{\alpha-2}{2}} (L-z)^{\frac{\alpha-2}{2}} dz, \quad 3 < \alpha < 5$$

$$(2.43b)$$

for a spherical wave, where

$$B(\alpha) = -\pi^3 \frac{1}{2\Gamma(\alpha/2)\cos(\pi\alpha/4)} \quad \text{and} \quad L_0 > (\lambda L)^{1/2} > l_0$$

Here, the source is located at $z = 0$ and the observation point is at L. For the case of constant β along the path, as might approximate a horizontal path, Equations 2.43a and 2.43b become

$$\sigma_X^2 = \frac{2}{\alpha} B(\alpha) A(\alpha) k^{\frac{6-\alpha}{2}} \beta L^{\frac{\alpha}{2}}, \quad 3 < \alpha < 5 \text{ (plane wave)} \quad (2.44a)$$

$$\sigma_X^2 = B(\alpha) \cdot \frac{\Gamma(\alpha/2)\Gamma(\alpha/2)}{\Gamma(\alpha)} A(\alpha) k^{\frac{6-\alpha}{2}} \beta L^{\frac{\alpha}{2}}, \quad 3 < \alpha < 5 \text{ (spherical wave)}$$

$$(2.44b)$$

For Kolmogorov turbulence ($\alpha = 11/3$) the log-amplitude variance (Rytov variance) has the classical form, with $\beta \equiv C_n^2$, and the expressions are given by [23]

$$\sigma_X^2(L) = 0.307 k^{7/6} L^{11/6} \beta \quad \text{(plane wave)} \quad (2.45a)$$

$$\sigma_X^2(L) = 0.124 k^{7/6} L^{11/6} \beta \quad \text{(spherical wave)} \quad (2.45b)$$

For N-O-K (non-Obukhov-Kolmogorov) turbulence, $\beta \neq C_n^2$ and needs to be recalculated. In Reference 47, an approach was proposed that relates the given spatial wavenumber of turbulence spectrum to the correlation length of intensity scintillations and generalized refractive index structure constant β. Thus, the equation for β is given by [47]

$$\beta = \frac{A(11/3)}{A(\alpha)} C_n^2 (\sqrt{k/L})^{\alpha - 11/3}, \qquad (2.46)$$

where $\beta \equiv \beta(z)$ and $\alpha \equiv \alpha(z)$ for vertical or slant-path propagation.

Following Reference 47, the generalized form of log-amplitude variance [for $C_X(L, 0) \equiv \sigma_X{}^2$) for plane-wave downlink propagation (source is located at L) is given by

$$\sigma_X^2 = 0.033 \cdot B(\alpha)(\sec \varphi)^{\frac{\alpha}{2}} k^{\frac{7}{6}} \int_0^L C_n^2(z) z^{\frac{5}{6}} \, dz, \quad 3 < \alpha < 5 \qquad (2.47)$$

For a spherical wave and uplink propagation (source is located at $z = 0$ and the receiver is at L), it yields

$$\sigma_X^2 = 0.033 \cdot B(\alpha)(\sec \varphi)^{\frac{\alpha}{2}} k^{\frac{7}{6}} \int_0^L C_n^2(z) \left(1 - \frac{z}{L}\right)^{\frac{\alpha-2}{2}} z^{\frac{5}{6}} \, dz, \quad 3 < \alpha < 5$$

$$\qquad (2.48)$$

These expressions are valid only for small apertures (point detector) with diameter $d < d_0 \approx (\lambda L)^{1/2} = (\lambda h \sec \varphi)^{1/2}$, where d_0 corresponds to the correlation length of intensity fluctuations when the propagation path length L satisfies the condition $l_0 < (\lambda L)^{1/2} < L_0$. The variance, σ_X^2, is dependent on the turbulence model assumed for the atmosphere. For α in vertical and slant-path propagation, a three-layer model as proposed in Reference 47 can be used with $\alpha \equiv \alpha(z)$. In the case of arbitrary power law exponent, the expressions are different from the "classical" (Kolmogorov) solution by the constants and path weighting for a spherical wave.

The log-amplitude variance as a function of α for Earth-space downlink and uplink propagation is shown in Figure 2.3. The scintillation effect is slightly reduced for higher power laws where $\alpha \rightarrow 5$.

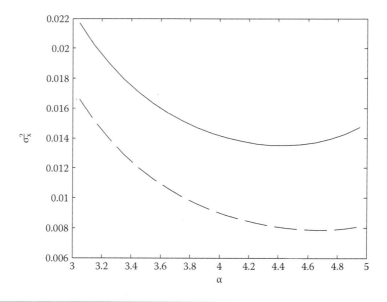

Figure 2.3 The log-amplitude variance as a function of α for Earth-space propagation: downlink (solid line) and uplink (dashed). $\lambda = 1.55$ µm; angle from zenith $\varphi = 0$, H-V 5/7 turbulence altitude model. (According to Reference 47.)

Here, the spectrum has a steeper slope (reduced energy of small eddies) and the phase effects become dominant. The refractive index fluctuation variance is influenced generally by large scale sizes.

It should be noted that the analysis and expressions presented here are related only to the inertial interval of turbulence spectrum. We do not describe the influences of small scales (dissipative interval) and anisotropic turbulence properties on wave propagation and scintillation statistics.

As already mentioned, the correlation length of the irradiance fluctuations is on the order of the first Fresnel zone, $\ell_F \approx \sqrt{L/k}$. However, measurements of the irradiance covariance function under strong fluctuation conditions reveal that the correlation length decreases with increasing values of the Rytov variance, σ_I^2. In other words, in the strong-fluctuation regime, the spatial coherence radius, ρ_0, of the wave determines the correlation length of irradiance fluctuations, and the scattering eddy characterizes the width of the residual tail, $x/\rho_0 k$.

In References 24–26, the theory developed in References 20–23 was modified for strong fluctuations, and it was shown that the smallest

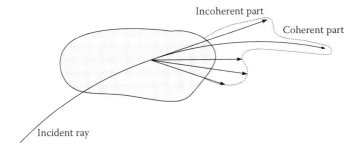

Figure 2.4 The total field pattern consisting of the coherent part (I_{co}) and incoherent part (I_{inc}). (From Reference 19.)

scales of irradiance fluctuations persist into the saturation regime. Kolmogorov theory assumes that turbulent eddies range in size from a macroscale to a microscale, forming a continuum of decreasing eddy sizes. These assumptions are based on recognizing that the distribution of refractive power among the turbulent eddy cells of a random medium is described by an inverse power of the physical size of the cell. As a coherent optical beam propagates through the random atmosphere, the wave is scattered by diffraction at the smallest of the turbulent cells (on the order of millimeters), creating a defocusing effect or diffractive scattering [19]. This kind of scattering is defined by the incoherent component of the total signal. Thus, because turbulence cells act as defocusing lenses, they decrease the amplitude of the wave by a significant amount for even short propagation distances. Schematically, such a situation is sketched in Figure 2.4, containing both components of the total field.

Refractive and diffractive scattering processes are compound mechanisms, and the total scattering process acts like modulation of small-scale fluctuations by large-scale fluctuations.

2.5.2 Beam Wander and Angle of Arrival of Optical Wave

The beam observed at some distance from its source will wander randomly in the plane transverse to the direction of propagation so that part or all of it will miss the receiving aperture or target. This is most likely for the case where the turbulence occurs close to the source.

In the absence of turbulence, a laser beam exiting from an aperture of diameter D would have an angular spread of order λ/D in the far field, where λ is the signal wavelength. Numerous turbulent eddies distributed across the beam cross-section cause many small angle scattering events to occur. This has the effect of spreading the beam over a wider cross-section, thus reducing the signal intensity at the receiver (or target).

In general, beam wander effects arise when the size of turbulence eddies is larger than the width of the beam that causes the wavefront tilt, whereas those eddies that are small compared with the beam diameter tend to widen the beam. The beam wander is important for beam pointing considerations, whereas the short-time widening is important for pulse propagation problems and high-energy laser systems. In particular, the wander and widening of a laser beam reduces its irradiance at the target/receiver plane and the precision of pointing at a target in military applications.

The beam wander can be characterized statistically by the variance of the beam centroid displacement (either magnitude of displacement or component along a single axis) at the target/receiver plane (or lateral position from initial direction of propagation). The beam centroid displacements or dancing in the focal plane of an optical system are associated with angle-of-arrival fluctuations of an optical wave in the plane of the receiver aperture. The kind of wavefront distortion that causes centroid dancing is wavefront tilt, when all rays across the aperture arrive at the same angle.

This happens when the part of the wavefront that reaches the aperture is relatively small compared to the size of the turbulence eddies near the receiver. Thus, the phase fluctuations are dominated by the low spatial frequencies of the turbulence spectrum. The low-frequency components are responsible for tilt effects, beam wander, and image dancing.

A short summary of approaches to estimate beam wander statistics and angle of arrival is presented here. These are important for solution of problems relating to optical communication.

As mentioned, turbulent cells larger than the beam diameter lead to refractive effects (beam wander) and can be characterized statistically by the variance of the random displacement of the beam centroid from its instantaneous position. Turbulence scale sizes smaller than the beam diameter produce diffractive effects that lead to beam widening.

Different approaches based on theory and experimental data have been proposed to describe beam displacement and wavefront angle of arrival for horizontal and slant-path propagation. Because the beam wander (or angle of arrival) is caused mostly by large-scale turbulence near the transmitter (receiver), the analysis is usually based on the geometric optics approximation (GOA) [23, 24], where diffraction effects are neglected. Under GOA, the Fresnel zone radius r_F corresponds to $r_F = [\lambda L]^{1/2} \ll l$, and $L \ll kl^2$, where l is the characteristic scale size of turbulent irregularities.

In this approximation the wave amplitude is still unchanged as the wavefront propagates through the optically inhomogeneous medium. The refractive index fluctuations cause only phase profile changing, as given by [21, 23, 42, 43]

$$\varphi(r,z) = -k \int_0^z n_1(r,z')dz' \qquad (2.49)$$

where $n_1 = n - <n>$ denotes the fluctuating part of the refractive index at a given point. Equation 2.49 expresses the additive principle of phase fluctuations. From Equation 2.49, it follows that under GOA the phase fluctuations at given points of wavefront (or aperture) are defined only by refractive index irregularities along the beam and are independent of characteristics of the propagating wave. The diffractive effects that cause the wave amplitude distortions are not taken into account.

To estimate wavefront distortions, the statistical properties of wave phase, $\Psi(r)$, which is the random function, need to be derived. The wavefront tilt with small angle $\delta\varphi$ yields a phase difference that is given approximately by $\Delta\Psi = \Psi(r_1) - \Psi(r_2) \approx k \cdot \rho \, \delta\varphi$, where $\rho = |r_1 - r_2|$ is the separation of two points in a transverse plane of wave propagation and $k = 2\pi/\lambda$.

By assuming that the mean $<\varphi> = 0$, the mean square of wavefront tilt fluctuations (or angle-of-arrival variance) can be expressed in the form [23, 24]

$$\langle\varphi^2\rangle \approx \frac{\langle(\Delta\psi)^2\rangle}{(k\rho)^2} = \frac{0.97 \cdot D_S(\rho)}{(k\rho)^2} \qquad (2.50)$$

where $D_S(\rho) = \langle[\psi(r)-\psi(r+\rho)]^2\rangle$ is the structure function of phase fluctuations (the variance of the phase difference) in the plane

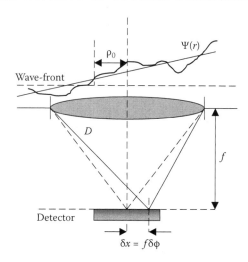

Figure 2.5 Wavefront tilt and centroid displacement on detector. The tilt, $\delta\varphi$ (in radians) of the wavefront is proportional to the displacement δx (in meters) of the centroid formed; f is the effective focal length (in meters).

transverse to the direction of wave propagation for the plane or spherical wave, and ρ corresponds to the receiver aperture diameter D in angle-of-arrival calculations. Thus, for an aperture of diameter D, the angle of arrival is directly related to the average tilt of the wavefront across the aperture and to the phase differences over the aperture (see Figure 2.5).

To estimate the effects of turbulence on optical wave propagation, the wave equation for random media needs to be solved. For weak, homogeneous, and isotropic turbulence, the method of small perturbations can be used to solve the wave propagation problem. The solution yields the phase structure function for a spherical or plane wave from a point source within the atmosphere, in the form [21–24, 42–44]

$$D_s(\rho,L) = 4\pi^2 k^2 L \int_0^1 \int_0^\infty \Phi_n(K) \cdot F(K,\rho,\xi) dK d\xi \qquad (2.51)$$

where ρ is the separation between points in the plane transverse to the direction of propagation (or in the receiver plane), $\Phi_n(K)$ is the three-dimensional spectral density of the index-of-refraction fluctuation, and the function $F(K,\rho,\xi)$ is the so-called spatial filter function that, under the GOA, has the form

$$F(K,\rho,\xi) = 2 \cdot [1 - J_0(K,\rho,\xi)] \cdot K \qquad (2.52)$$

where J_0 is the zero-order Bessel function of the first kind. For different values of ρ, the function $F(K,\rho,\xi)$ derives the contribution of different spectral scale sizes of refractive index fluctuations to the phase structure function. It is assumed that beam intensity fluctuations can be ignored.

The phase structure function of a spherical wave for the Obukhov-Kolmogorov spectrum of refractive index fluctuations and slant-path propagation takes the form [23]

$$D_S(\rho) = const \cdot \sec\theta \cdot k^2 \rho^{5/3} \overline{C}_n^2, \qquad (2.53)$$

where $\overline{C}_n^2 = \int_0^L C_n^2(z)(z/L)^{5/3} dz$ is the path-integrated value of the refractive index structure parameter and θ is the zenith angle. Here, the integration is performed downward over the slant path from the source of the optical wave (at altitude or distance L) to the receiver. For upward integration, the path-integrated value of C_n^2 can be replaced by $\overline{C}_n^2 = \int_0^L C_n^2(z)(1 - z/L)^{5/3} dz$.

The mean square angle-of-arrival fluctuation is found to be (one-directional variance)

$$\langle \varphi^2 \rangle = 0.97 \cdot \frac{const \cdot \overline{C}_n^2}{D^{1/3}}, \quad rad^2 \qquad (2.54)$$

where $const = 1.46$ for the case of $l_0 \ll D \ll (\lambda L)^{1/2}$ and $const = 2.914$ for $(\lambda L)^{1/2} \ll D \ll L_0$, and D is the receiver diameter. Here the point source is assumed. The theoretical results were verified experimentally at different distances for horizontal propagation [23, 48].

The beam wander corresponds to image motion where the system's focal length converts angular deviations into position deviations (see Figure 2.5). Thus, beam wander can be modeled as if it arises from a random tilt angle at the transmitter plane, similar to angle-of-arrival fluctuations of a reciprocal propagating wave with the receiver diameter replaced by the transmitter beam diameter. By using this principle, the beam wander statistics can be obtained on the basis of Equation 2.54 by replacing the receiver diameter D by the transmitter diameter D_0.

The variance of beam centroid displacements (beam wander) at a given distance L can be represented as

$$\langle \rho_C^2 \rangle = \frac{D_S(z, D_0) \cdot L^2}{k^2 \cdot D_0^2}, \quad m^2 \tag{2.55}$$

For horizontal propagation path, the solution takes the form

$$\langle \rho_C^2 \rangle \approx 1.1 \cdot L^3 C_n^2 D_0^{-1/3}, \, (\lambda L)^{1/2} \ll D_0 \ll L_0 \tag{2.56}$$

where the phase structure function of a spherical wave in the form of Equation 2.53 was used.

The centroid position can be converted to the wavefront tilt (or angle of arrival). Figure 2.5 illustrates the relationship of an incident tilted wavefront and spot position on the sensor for the 1D case. The receiver effective focal length, f, converts wavefront slope into a position in the focal plane, $x = f\varphi$.

The beam wander and wave angle-of-arrival variance in the plane of the receiver aperture is directly related to the path-integrated value of C_n^2 as

$$\langle \varphi^2 \rangle \sim \bar{C}_n^2 = \int_0^L C_n^2(z) \cdot Q(z, L) dz, \tag{2.57}$$

where $Q(z,L)$ is the optical weighting function for the path and L is the path length. This function describes the relative effectiveness of each portion of the propagation path for producing wavefront tilt (or angle-of-arrival fluctuation).

The phase structure function can be represented in terms of transverse atmospheric coherence length (or turbulence coherence diameter):

$$D_S(\rho) = 2[\rho/\rho_0]^\alpha, \quad l_0 \ll \rho \ll L_0 \tag{2.58}$$

where $\alpha = 5/3$ for Kolmogorov turbulence. The parameter ρ_0 characterizes the spatial scale of phase fluctuations; i.e., the correlation length of phase fluctuations.

Following Equation 2.53, under the Obukhov-Kolmogorov turbulence model, the turbulence coherence diameter is given by

$$\rho_0 = \left[1.457 \cdot \sec\theta \cdot k^2 \cdot \bar{C}_n^2 \right]^{-3/5}. \tag{2.59}$$

As follows from Equation 2.58, the fluctuations of phase difference across the wavefront decrease when the transverse distance ρ starts to be smaller than ρ_0. In imaging applications, the atmospheric phase fluctuations are insignificant for receiver aperture diameter $D < \rho_0$.

For the aperture $D \sim \rho_0$, the wavefront can be approximated as linear, corresponding to local wavefront tilt (see Figure 2.5). The wavefront tilt appears as image jitter or dancing. If the receiver aperture is small, i.e., $D \ll \rho_0$, the magnitude of jitter lies inside the width of diffraction circle and the wavefront tilt (or jitter) influence is negligible.

In the case of $D > \rho_0$, the wavefront curvature/distortion must be taken into account in imaging applications (see Figure 2.5); the wavefront tilt in different regions of the aperture will be uncorrelated, adding blur to the image and leading to a reduction in image motion. In addition, a nonuniform irradiance distribution will give unequal weights to tilts in different regions of the aperture. As the aperture dimensions exceed the irradiance correlation distance (i.e., $D \gg [\lambda L]^{1/2}$), this effect will diminish because the average irradiance pattern predominates over the fluctuating part. The resultant image for a receiver of diameter $D > \rho_0 > [\lambda L]^{1/2}$ will be blurred and exhibit a center-of-gravity displacement.

Another derivation of the correlation scale of phase fluctuations is in terms of the Fried parameter or Fried's coherence diameter [23, 24, 42–44]:

$$D_S(\rho) = 6.88[\rho/r_0]^{5/3}, \quad l_0 \ll \rho \ll L_0 \tag{2.60}$$

The Fried coherence diameter for the Kolmogorov turbulence statistics is given by

$$r_0 = \left[0.423(2\pi/\lambda)^2 \sec\theta \cdot \overline{C}_n^2 \right]^{-3/5} \tag{2.61}$$

The turbulence coherence diameter is related to the Fried parameter by $r_0 = 2.1\,\rho_0$ [29, 67]. The concept of the atmospheric turbulence coherence length is useful in understanding the limited resolution achievable in imagery gathered through the atmosphere. Optical turbulence acts like a low-pass filter on an optical system by filtering out the high spatial frequencies.

The resolving power is limited by the optics when the diameter, D, is smaller than r_0 and limited by the atmosphere when D is larger than r_0. The coherence diameter, r_0, is also called the "seeing parameter" because large values of r_0 mean "good seeing." At optical wavelengths, diffractive effects can start to be significant even for relatively short path lengths (200–1000 m). For larger distances of wave propagation in a turbulent atmosphere, the additive principle (Equation 2.49) of phase fluctuations is not fulfilled. In this case the method of small perturbations is used, which includes the diffractive effects.

In earlier works, the GOA approach to calculate the wander of a single ray was used [49, 50], and equivalent calculations of Gaussian beam wander based on the gradient of the refractive index along the beam were performed using a Huygens-Fresnel formulation [51]. However, effects of the finite beam size were not included.

The Huygens-Fresnel formulation was a basis for other treatments restricted to the finite focused beam case [52, 53], and reasonable agreement with focused beam experimental data has been obtained [54]. For Gaussian beam propagation, a typical focused beam result is given in a review paper by Fante [44]. The finite beam effects were described by Chiba [55] for a collimated or nearly collimated beam. Churnside and Lataitis [56] developed an analytical expression for the beam wander variance using a GOA applied to a finite, uniform beam propagating through weak refractive turbulence with the Kolmogorov spectrum.

A more comprehensive analysis of beam wander statistics that includes diffraction effects and the effects of scale sizes outside the inertial sub-range has been developed using the Markov approximation [23, 44, 57]. It was assumed that the index of refraction fluctuation is a delta function correlated in the direction of propagation. To include outer scale effects, the von Karman (Equation 1.36) and exponential (Equation 1.37) forms of the refractive index fluctuation spectrum were used (see Chapter 1). Klyatskin and Kon [58] developed an expression for the total beam wander variance that is valid under weak and strong irradiance fluctuations. Mironov and Nosov [48] developed separate asymptotic relations for the beam centroid variance and compared it with various experimental data for a focused beam.

Tofsted [59] presented the approximated expressions for the beam wander and angle-of-arrival variances of a Gaussian beam and a uniform-intensity cross-section beam. A similar result was obtained earlier by Cook [60], who used the optical analog of Ehrenfest's theorem to obtain the variance of beam centroid displacement and angle of arrival of a finite beam and applied it to the particular cases of focused and collimated Gaussian beams. A modified von Karman spectrum of refractive index to account for finite outer scale was used.

Andrews and Phillips [24] presented models for beam centroid variance for the case of a Gaussian beam wave where the beam characteristics (collimated or focused) and the outer turbulence scale variability were taken into account. The models have been generalized also for the cases of moderate-to-strong irradiance fluctuations. On the basis of scalar parabolic equation approximation for propagation of optical radiation, an expression for lateral beam centroid displacement at a given distance can be described by the vector [48, 61, 62]

$$\rho_{LB}(L) = \frac{1}{2P_0} \int\limits_0^L (L-z)dz \int I(z,R) \nabla_R n_1(z,R) d^2R \quad (2.62)$$

where $\nabla_R n_1$ is the transverse gradient of the atmospheric refractive index, $I(z, R)$ is the beam intensity distribution at a distance z from the transmitter, and $P_0 = \int I(L,R)d^2R$ is the full power of the beam. R is the coordinate in the plane transverse to the direction of propagation.

Thus, the position of the beam center of gravity at distance L is defined by the transverse gradient of the refractive index (or, in the general case, the dielectric permittivity of the medium), which is summed in volume between the transmitter plane and some plane at distance L with the kernel that equals the beam intensity distribution.

Using this formulation, the expression for a mean square of the location vector of the beam center of gravity can be written with the Markovian process approximation (i.e., the correlation function of the refractive index fluctuations is assumed to be delta-correlated in the direction of propagation) in the form [24, 48, 58, 59, 63]

$$\left\langle \varphi_{LB}^2 \right\rangle = 4\pi^2 L \int\limits_0^1 d\xi (1-\xi)^2 \int\limits_0^\infty dK \Phi_n(K,L\xi) K^3 \cdot \exp[-V(K,a,L\xi)]$$
$$(2.63)$$

where $z = L\xi$ is the distance along the path, $\Phi_n(K, z)$ is the three-dimensional spectrum of the field n_1 depending on the two-dimensional wavenumber vector $K = (K_x, K_y, K_z = 0)$ [i.e., the power spectrum is reduced to $\Phi_n(K_x, K_y, 0, z)$], and the second integral has the form [48] $\int d^2 K \Phi_n(K, L\xi) K^2 \exp[-V]$, where $\exp[-V]$ is the normalized Fourier transform of the instantaneous beam irradiance profile across a plane at position z, and $V(K, a, z)$ can be described as an attenuation factor or filter function related to the effectiveness of different turbulence scale sizes to create tilt, where a is the beam parameter.

In the integral solution (Equation 2.62), the field n_1 was assumed to be statistically homogeneous and Gaussian, and the second moment of radiation intensity has been approximated by the mean intensity squared. This approximation is valid as long as the beam does not become highly scintillated.

It was assumed also that functions $I(z, R)$ and $\Phi_n(K, z)$ are isotropic and the average intensity distribution $<I(L,R)>$ in the plane $z = L$ has a Gaussian profile. The attenuation factor $V(K, a, z)$ has a form [64]

$$V = \frac{K^2[a_0 q(\xi)]^2}{2} \qquad (2.64)$$

where

$$q^2(\xi) = \left[\Theta_0^2 + \Omega^{-2} + \Omega^{-2}\left(\frac{1}{2}D_S(\rho)\right)^{6/5} \right] \qquad (2.65)$$

where $\Theta_0 = 1 - \xi L/F$; $\Omega = ka_0^2/(L\xi)$ is the Fresnel number; a_0 and F are the e^{-1}-intensity radius of the Gaussian beam at the source and the focal range of the beam at the source, respectively; and $D_S(\rho)$ ($\rho = 2a_0$) is the phase structure function for a spherical or plane wave that describes the phase fluctuation in a plane transverse to the direction of wave propagation.

By using the refractive index spectrum in the Kolmogorov form ($\Phi_n(K) = 0.033C_n^2 K^{-11/3}$), one can approximate the phase structure function $D_S(\rho)$ by

$$D(\rho) = 2(\rho/\rho_0)^{5/3} \approx 2(\rho/\rho_0)^2 \qquad (2.66)$$

Here ρ must satisfy $l_0 \ll \rho \ll L_0$, where l_0 and L_0 are the turbulence inner and outer scales, respectively. Equation 2.66 is the quadratic structure function approximation. Kon et al. [65] found that this approximation gives results within 4% difference of those obtained with the 5/3 power law expression in Equation 2.66.

The quantity ρ_0 is the long-term lateral coherence length of a spherical or plane wave. For a spherical wave, it is given by [44]

$$\rho_0(L) = \left[1.46 \cdot k^2 \int_0^L C_n^2(z)(z/L)^{5/3} dz \right]^{-3/5} \quad (2.67)$$

The short-term coherence length, ρ_0^{ST}, can be approximated as [66]

$$\rho_0^{ST} \cong \rho_0 \cdot [1 + 0.37 \cdot (\rho_0/a_0)^{1/3}] \quad (2.68)$$

Using the phase structure function given by Equation 2.66, the expression for q can be presented as

$$q^2(\xi) = (1 - \xi L/F)^2 + \left(\xi L/ka_0^2 \right)^2 + 4(\xi L/a_0 k \rho_0)^2 \quad (2.69)$$

The term $[a_0 q(\xi)]^2$ in Equation 2.64 corresponds to the mean square radius of the beam spot; i.e., the radius at which the long- and short-term average intensity distribution is reduced by a factor of e^{-1} from its maximum value. The first two terms in Equation 2.69 represent the beam spreading in a vacuum due to diffraction, and the last term represents the additional spread due to turbulence.

Using the Kolmogorov spectrum and the quadratic approximation to the phase structure function (Equation 2.66), the solution (Equation 2.63) for total beam centroid variance at a given distance can be represented as follows:

$$\left\langle \varphi_{LB}^2 \right\rangle = C_1 \cdot L a_0^{-1/3} \int_0^1 d\xi (1 - \xi)^2 q^{-1/3}(\xi) C_n^2(L\xi) \quad (2.70)$$

where $C_1 = 4\pi^2 \, 0.033 \, \Gamma(1/6) 2^{-5/6} \approx 4.07$, and $\Gamma(*)$ is the gamma function.

Although often omitted in theoretical analyses, the presence of a finite outer scale can affect the amount of beam wander that actually occurs. The outer scale effects can be introduced through use of different spectral models.

The *modified von Karman* (Equation 1.36) and *exponential* (Equation 1.37) isotropic forms of spectra (see Chapter 1), which take into account deviations from a power law in the region of a turbulence outer and inner scale, can be used [24, 64]. As was mentioned in Chapter 1, these forms of the turbulence spectrum are used for computational reasons and are not based on physical models. The effects of the turbulence outer scale tend to decrease the mean square of beam displacements. In the case of vertical or slant-path propagation with small angle from zenith, the inner scale effects can be assumed to be negligible.

By using the exponential turbulence spectrum and Equations 2.63–2.66, the solution for $\langle \varphi_{LB}^2 \rangle$ is represented by

$$\left\langle \varphi_{LB}^2 \right\rangle = C_1 \cdot L a_0^{-1/3} \int_0^1 d\xi (1-\xi)^2$$

$$\times \left[q^{-1/3}(\xi) - \left(q^2(\xi) + \frac{2}{a_0^2 K_0^2(\xi L)} \right)^{-1/6} \right] C_n^2(L\xi) \tag{2.71}$$

If we assume that the outer turbulence scale is large $[a_0 q(\xi) \ll L_0]$, then Equation 2.71 reduces to Equation 2.70.

In Reference 24, Andrews and Phillips presented a general expression for the variance of beam centroid displacement under weak irradiance fluctuations. A Gaussian beam wave was assumed, and the combination of a Markov approximation and GOA with exponential turbulence spectrum (Equation 1.37) was used. The expression is given as follows:

$$\left\langle \varphi_{LB}^2 \right\rangle = C_3 \cdot L \cdot a_0^{-1/3} \int_0^1 C_n^2(\xi L)(1-\xi)^2$$

$$\times \left\{ \frac{1}{|\Theta_0|^{-1/3}} - \left\{ \frac{K_0^2(\xi L)a_0^2}{1 + K_0^2(\xi L)a_0^2 \Theta_0^2} \right\}^{-1/6} \right\} d\xi \tag{2.72}$$

where $C_3 = 7.25$; a_0 and $\Theta_0 = 1 - \xi L/F$ denotes the Gaussian beam radius in the plane of emitting aperture and the input plane beam parameter, respectively; and F is the phase front radius of curvature. In the case of a collimated beam, $\Theta_0 = 1$. The final expression is nearly the same as Equation 2.71 except for the constant C, and the beam widening effects due to diffraction and turbulence are not included.

For the case of an infinite outer scale ($K_0 = 0$), Equation 2.72 gives the same result as that obtained by Churnside and Lataitis [56] for the finite uniform beam using the GOA except for a difference in the constant.

Cook [60] used the optical analog of Ehrenfest's theorem to obtain the variance of beam centroid displacement and angle of arrival of a finite beam and applied them to the particular cases of focused and collimated Gaussian beams. Following Cook, the variance of linear beam displacement is given by

$$\langle \rho^2 \rangle = 4 \cdot \int_0^L (L-z)^2 D_\mu(z) dz \tag{2.73}$$

where

$$D_\mu(z) = -\frac{1}{4} \int_{-\infty}^{\infty} \nabla_T^2 B_\mu(0,z,L) dz \tag{2.74}$$

is the ray-diffusion constant for the refractive index, $\nabla_T^2 = \delta^2/\delta x^2 + \delta^2/\delta y^2$ is the transverse Laplacian, and B_μ is the correlation function of the refractive index.

The diffusion constant was calculated by using the modified von Karman spectrum to account for finite outer scale and is given by

$$D_\mu(z) = 0.033\pi^2 C_n^2(z) \int_0^\infty \frac{\exp\left\{-K^2\left(a^2(z) + 4l_0^2\right)/4\right\} K^3}{\left[K^2 + L_0^{-2}\right]^{11/6}} dK \tag{2.75}$$

where $a(z)$ is the beam size. In general, $a(z)$ can be derived as $a^2(z) = (a_0 q(z))^2$ from Equation 2.65 (Gaussian beam) or as $a = a_0|\Theta_0|$ under GOA, and $L_0 = L_0(z)$ is the function of height.

Introducing the variables $Q^2 = (a^2 + 4\,l_0^2)/4$, $A^2 = (Q/L_0)^2$, and $u = Q \cdot K$, we can write Equation 2.75 in the form

$$D_\mu = 0.033\pi^2 C_n^2 \cdot Q^{-1/3} \cdot G(A) \tag{2.76}$$

where $G(A)$ is given by

$$G(A) = \int_0^\infty \frac{\exp(-u^2)}{[u^2 + A^2]^{11/6}} u^3 du \tag{2.77}$$

The integral solution gives

$$
G(A) = \frac{1}{2} A^{1/3} \frac{\Gamma(-1/6)}{\Gamma(11/6)} {}_1F_1\left(2; \frac{7}{6}; A^2\right)
$$

$$
+ \frac{1}{2} \cdot \Gamma(1/6) \, {}_1F_1\left(\frac{11}{6}, \frac{5}{6}; A^2\right) \tag{2.78}
$$

$$
\approx -3.6 \cdot A^{1/3} \, {}_1F_1\left(2; \frac{7}{6}; A^2\right) + 2.78 \, {}_1F_1\left(\frac{11}{6}, \frac{5}{6}; A^2\right)
$$

where $\Gamma(*)$ is the gamma function and ${}_1F_1(a,b,x)$ is the confluent hypergeometric function.

The parameter A can be approximated as $A \sim a(z)/L_0(z)$, if the beam spot size is much larger than the inner scale ($a^2(z) \gg 4\, l_0^2$). Thus, the final result can be represented as

$$
\langle \rho^2 \rangle = 1.3 \cdot \int_0^L (L-z)^2 C_n^2(z) Q^{-1/3}(z) G(A) dz \tag{2.79}
$$

where $Q(z) = (a^2(z) + 4\, l_0^2)^{1/2}/2$.

Tofsted [59] obtained similar results using the modified von Karman spectrum, the Markov approximation, and the quadratic phase structure function (Equation 2.66). The new variable $Q(z)$ in Equation 2.79 was introduced in the form [59]

$$
Q^2 = \frac{D_0^2}{8}\left(1 - \frac{z}{F}\right)^2 + \frac{1}{8}\left(\frac{4z}{k}\right)^2 \left(\frac{1}{D_0^2} + \frac{\pi}{\rho_0^2}\frac{z}{L}\right) + K_m^{-2} \tag{2.80}
$$

where D_0 is the diameter of the Gaussian beam at the source for decay in $\exp(-1)$ of the beam intensity and $A = Q^2/L_0^2$.

An approximation to the exact form for $G(A)$ was given by [59]

$$
G(A) \approx \begin{cases} 3.626 - 4.69 A^{1/6} + 1.9 A^{0.82} \exp(-0.6A) & 0 \le A < 0.4 \\ 11.43 \exp(-4.402 A^{0.2268}) - 0.107 \exp(-6A) & 0.4 \le A < 19 \\ 0.6514 A^{-11/6} & 19 \le A \end{cases}
$$

$$
\tag{2.81}
$$

For pulse-beam propagation and long distances, the short-term beam spreading (short-term coherence length) should be included.

In Reference 67 the exponential spectrum (see Chapter 1) for phase structure function calculations (Equation 2.66) to account for finite outer scale was used, and the approximated expression for the path-integrated value of C_n^2 in Equation 2.54 was found to be

$$\bar{C}_n^2 \cong \int_0^L dz C_n^2(z) \cdot \left\{ \left(1 - \frac{z}{L}\right)^{5/3} + \frac{3}{(K_0(z)D_0)^{5/3}} \cdot \left[1 - {}_1F_1\left(-\frac{5}{6}, 1, -b\right)\right] \right\}$$

(2.82)

where

$$b = \frac{(D_0 K_0(z))^2}{4} \cdot \left(1 - \frac{z}{L}\right)^2.$$

For slant-path propagation, the dependence on zenith angle θ ($<\varphi^2> \sim \sec\theta$) should be included. For uplink slant-path propagation and propagation inside the optical active turbulence layer, some form of saturation of variance of angular displacements with path length increase can be observed. It follows from the fact that phase effects (wavefront tilt) are dominated by strong turbulence that occurs near the receiving or transmitting aperture because of their low elevations and in the atmospheric boundary layer.

The expressions described here for variance of beam displacements correspond to the general case of beam propagation. In the case of nearly vertical propagation and not long distances (10–15 km), the effects of turbulence-induced beam widening can be neglected and Equations 2.65 and 2.69 reduce to

$$q^2(\xi) = \Theta_0^2 + \Omega^{-2}$$

(2.83)

where $\Theta_0 = 1$ for a collimated beam. Here it was taken into account that the initial beam radius is not wide; i.e., $a_0 \le [\lambda L]^{1/2}$ and $\rho_0 \gg [\lambda L]^{1/2}$.

In Figure 2.6 are shown angular beam displacement variances (μrad^2) at 10 km altitude for various initial diameters of collimated beams. The finite outer scale described by Equation 1.42 with $K_0(z) = 2\pi/L_0(z)$ is included in the calculations.

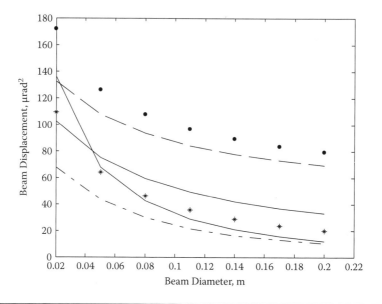

Figure 2.6 Angular beam displacements (μrad^2) at 10 km altitude versus the initial size of the collimated beam (according to Reference 67), calculated using Equation 2.70 (dashed line), Equation 2.71 (dash-dot line), Equation 2.72 (solid line), Equation 2.79 (dots), Equation 2.54 (bold dots), and Equation 2.63 (stars); the C_n^2 profile is given by Equation 2.82. (From A. Zilberman and N. S. Kopeika, "Laser beam wander in the atmosphere: Implications for optical turbulence vertical profile sensing with imaging LIDAR," *J. Appl. Remote Sensing*, vol. 2, 023540 (7 October 2008); DOI:10.1117/1.3008058. SPIE. With Permission.)

The angular displacement is greatest for smaller beams and decreases with increasing beam radius. Also, the presence of a finite outer scale reduces the amount of beam wander. The deviations in the results follow from the differences in constants.

Note that in Equation 2.54 the receiver diameter D was replaced by transmitter diameter $2a_0$, and the path-integrated value of C_n^2 is in the form $\bar{C}_n^2 = \int_0^L C_n^2(z)(1 - z/L)^2 dz$. Various models developed for beam wander and/or angle-of-arrival (AOA) analysis generally have the same functional form but different scaling constants.

The scaling constant in Equation 2.72 is different from that in Equations 2.70 and 2.71 by the factor 0.56; i.e., $C_1 = 0.56 C_3$. Following Reference 24, the difference arises because beam wander variance includes both centroid and hot spot displacement, where the hot spot is defined as the point of maximum irradiance within the beam profile.

As described in Chapter 1, recent evidence has shown significant deviations from the Kolmogorov model in certain portions of the atmosphere. The power spectrum of turbulence in the troposphere and stratosphere may exhibit non-Kolmogorov statistics.

Using Mellin transforms, Stribling et al. [46] obtained a general form of the phase structure function for a spherical wave with an arbitrary power law exponent α as follows:

$$D_S(\rho) = -2^{4-\alpha}\pi^2 \left(\frac{2\pi}{\lambda}\right)^2 A(\alpha)\frac{\Gamma\left(\frac{2-\alpha}{2}\right)}{\Gamma(\alpha/2)}\rho^{\alpha-2}\sec\theta\int_0^L \beta(z)(z/L)^{\alpha-2}dz$$

(2.84a)

where $3 < \alpha < 4$. When α is larger than 4, one needs to consider effects of the outer scale of turbulence.

The phase structure function can be represented in general form in terms of the generalized Fried coherence diameter as [68]

$$D_S(\rho) = C_1 \cdot \left(\frac{\rho}{r_0}\right)^{\alpha-2}, \quad 3 < \alpha < 4, \qquad (2.84b)$$

where

$$C_1 = 2 \cdot \left[\frac{8}{\alpha-2}\Gamma\left(\frac{2}{\alpha-2}\right)\right]^{\frac{(\alpha-2)}{2}}.$$

The generalized Fried coherence diameter for a spherical wave is given by [68]

$$r_0 = \left[C_2(2\pi/\lambda)^2\sec\theta\int_0^L dz\beta(z)(z/L)^{\alpha-2}\right]^{-1/(\alpha-2)} \quad 3 < \alpha < 4, \quad (2.85)$$

where

$$C_2 = \frac{\pi^{1/2}\Gamma\left(\frac{2-\alpha}{2}\right)}{2\Gamma\left(\frac{3-\alpha}{2}\right)\cdot\left[\frac{8}{\alpha-2}\Gamma\left(\frac{2}{\alpha-2}\right)\right]^{(\alpha-2/2)}}, \quad 3 < \alpha < 4,$$

It is customary to make the distinction between the short- and long-term effects. If the target is placed at a distance z, a widened laser beam spot of radius ρ_S deflected by a distance ρ_C from its line-of-sight (LOS) propagation is observed on the target (for short exposure of the beam on the target). If exposure time on the target is much longer than $\Delta t = D/V$ [V is the transverse velocity of the turbulent eddies and $D = a_0 q(\xi)$ is the beam size as given by Equation 2.69], the wander of the center of gravity of the beam ρ_C in addition to the short-term spread ρ_S contributes to the beam size on the target plane, and the broadened beam spot is observed with root mean square (RMS) radius ρ_L, given by [24, 44]

$$\rho_L = [\langle \rho_S^2 \rangle + \langle \rho_C^2 \rangle]^{1/2}. \tag{2.86}$$

Thus, the long-term beam radius ρ_L comprises a widened beam of RMS radius $[\langle \rho_S^2 \rangle]^{1/2}$ that wanders about with an RMS radius $[\langle \rho_C^2 \rangle]^{1/2}$.

Assuming a Gaussian beam intensity profile, the irradiance distribution on the target plane at the distance z is given by

$$I(z,r) = \frac{a_0^2}{W_e^2} \exp\left(-\frac{2r^2}{W_e^2} \right), \tag{2.87}$$

where a_0 is the initial beam radius and $W_e = a_0 q(\xi)$ is the effective beam spot size on the target at distance z from the transmitter. The beam width a_0 is the beam radius at the Gaussian $1/e$ point ($1/e$ of the maximum amplitude). The Gaussian beam width is related to the aperture diameter D_0 by $a_0 = D_0/2^{1/2}$. The effective beam spot size derived from turbulence effects (the short-term beam radius) is given by Equation 2.69.

2.6 Measurements of Atmospheric Turbulence

Experimental data has been collected and modeled in experiments the world over, including References 70–79. These have included observations in different climates and environments, because the dependence of C_n^2 on weather is of great interest [29, 69–75, 77, 78]. In general, C_n^2 increases with heating of the ground and thus is maximum around midday. At night, ground temperature is colder than air temperature, thus producing a temperature gradient opposite to that in daytime.

Temperature gradient approaches zero near sunrise and sunset, and that is when C_n^2 exhibits minimum value. However, over water there are no diurnal minima, usually because the very large heat capacitance of the sea keeps surface temperature fairly constant often over several

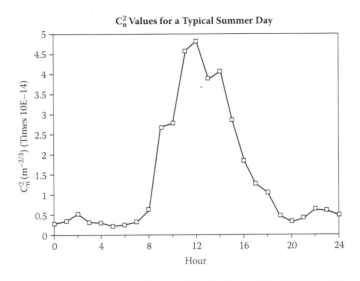

Figure 2.7 C_n^2 values for a typical summer day as determined from He-Ne beam-of-arrival fluctuations. (From N. S. Kopeika, *A System Engineering Approach to Imaging*, SPIE Optical Engineering Press, Bellingham, WA, 1998. With Permission.)

weeks, rather than changing between day and night as land temperature does. Parameter C_n^2 decreases with elevation and with increasing cloud cover, because clouds cause land temperature to diminish. It also decreases generally with distance from the equator, because surface heating is diminished, depending on season. Figure 2.7 shows an example of values of C_n^2 during the course of a typical summer day, at about a 15 m elevation [29, 72]. The diurnal minima near sunrise and sunset are quite obvious, as well as the maximum around midday. The sudden drops in C_n^2 observed at about 10:30 a.m. and 1:00 p.m., and midnight are a result of passing clouds.

One method to obtain a path average of C_n^2 is to measure angle-of-arrival fluctuations. Such fluctuations of an optical wave in the plane of the receiver aperture were associated with image dancing in the focal plane of an imaging system. To derive the path average mean-square angle of arrival, $\langle \alpha^2 \rangle$, a retrieving technique based on center-of-gravity variance calculation is used for a set of images. For a spherical wave, it is given by [29, 69]

$$\langle \alpha^2 \rangle = 2.914 \cdot D^{-1/3} \int_0^L C_n^2(z)(z/L)^{5/3} \, dz \qquad (2.88)$$

for $\sqrt{\lambda L} \ll D \ll L_0$, where D is the aperture diameter and L is the length of the atmospheric path.

After integration, Equation 2.88 reduces to

$$\langle \alpha^2 \rangle = 1.09 C_{neq}^2 L D^{-1/3} \qquad (2.89)$$

where L is path length, D is aperture diameter, and C_{neq}^2 is the path-integrated average value of C_n^2. This method is often preferred over scintillation measurement in order to avoid saturation regimes [69].

2.7 Modeling of Atmospheric Optical Turbulence

2.7.1 Analytical Models

As was mentioned in Chapter 1, many analytical physical models have been developed to describe atmospheric turbulence effects, such as von Karman [24, 26], Beckman [21, 24], Monin-Obukhov [70], Thiermann [71], and so on [72–74]. The Kolmogorov model [75] was chosen because of its simplicity and straight connection to the physical parameters of the atmospheric channel. Using the Kolmogorov energy cascade theory of turbulence [21, 23, 24, 69] (see also Chapter 1), we can divide the turbulence scale into three regions by two scale sizes: the outer scale of turbulence, L_0, and the inner scale of turbulence, l_0. As was mentioned in Chapter 1, the value of L_0 may vary widely. Near the ground, it usually equals the height over the land. In the free atmosphere, it may range from 10 to 100 m or more, changing significantly in different regions of space, whereas l_0 is usually 1 to 10 mm.

As was mentioned also in Chapter 1, the physical phenomena, caused by atmospheric turbulence, may be divided into three cases [23, 69, 75]. The first is $L_0 < l$ and $\kappa < 2\pi/L_0$, where l is the size of the turbulence eddies and κ is the spatial wavenumber of eddies. In this case, large-scale atmospheric features, such as wind, form turbulence eddies of such dimensions. Here, energy is input to the turbulence system from the thermal and kinetic energy of the atmosphere [23, 69]. This process depends on climatic conditions mainly. The second case is $l_0 < l < L_0$ and $2\pi/L_0 < \kappa < 2\pi/l_0$. Here eddies, formed in the input range, are unstable and fragment into smaller eddies [24, 75, 76], causing energy to be distributed from small to large turbulence wavenumbers, whereas energy loss is minor. The third case is $l < l_0$ and $2\pi/l_0 < \kappa$. Here, turbulence energy, which was transferred through the inertial sub-range, is transferred through cracks in the small eddies. The corresponding analytical models giving spectral characteristics of the atmospheric turbulence for different vertical and horizontal traces were described in Chapter 1.

2.7.2 Empirical and Semi-Empirical Models

As for the empirical and semi-empirical models of atmospheric turbulence, we consider in this section those that describe atmospheric elevations near the ground surface. Higher elevations can be very important for optical communication between skyscrapers. The corresponding models for vertical traces were discussed in Chapter 1. Investigations have shown that for the near-ground horizontal atmospheric links, C_n^2 depends on elevation, surface and meteorological conditions, and latitudes, and generally falls in the range 10^{-15} to 10^{-13} m$^{-2/3}$ [72, 77–83]. At heights of 1–2 km, some empirical models based on measurements carried out at different latitudes can also be useful to estimate C_n^2. Thus, one of the most commonly used models of turbulence strength vertical profile is the Hufnagel-Valley (H-V) model [79]. This single-parameter model is determined from upper-altitude winds. However, this model is developed only for a mid-latitude subtropical atmosphere and its performance can be poor for other sites. Fortunately, a Middle East refractive index structure constant vertical profile model has been developed [79]. The model form is similar to a generalized H-V model, but with specific turbulence layers. Such distinctly layered turbulence structure has been confirmed in many measurements performed in the past employing passive and active methods, and through *in situ* observations [81–83].

We will discuss the data predicted experimentally and theoretically by the corresponding models and introduce a new model, with two "practical" variations, as an extension of the macroscale meteorology model [29, 72, 86], for prediction of C_n^2 in the turbulent atmosphere above the midland coastal zones, where this macroscale model has essential limitations.

2.7.2.1 Concept and Applications of Thiermann (MOS) Model Refractive index fluctuations are a consequence of air temperature fluctuations, which are characterized by the temperature structure parameter C_T^2 (in $K^2 m^{-2/3}$). Ignoring minor wavelength and humidity dependences, C_n^2 can often but not always [29] be expressed in the following form over land [84, 87]:

$$C_n^2 = \left(79 \cdot 10^{-6} \cdot \frac{P}{T^2} \right)^2 C_T^2 \qquad (2.90)$$

where P (in millibars) is the air pressure and T (in degrees Kelvin) is the ambient temperature. Equation 2.90 is valid in the visible and infrared wavebands from approximately 0.5 to 10 µm. A model of atmospheric turbulence developed by Thiermann [71] provides C_T^2. Here Equation 2.90 is used to determine the values of the refractive index structure parameter. The vertical profiles of C_T^2 according to Reference 88 are

$$C_T^2 = 4\beta \frac{T_\otimes^2}{(kz)^{2/3}} \left[1 + 7\frac{z}{L_\otimes} + 20\left(\frac{z}{L_\otimes}\right)^2 \right]^{-1/3} \qquad (2.91a)$$

for the stable case ($z/L_\otimes > 0$), where air is warmer than the ground, and

$$C_T^2 = 4\beta \frac{T_\otimes^2}{(kz)^{2/3}} \left[1 - 7\frac{z}{L_\otimes} + 75\left(\frac{z}{L_\otimes}\right)^2 \right]^{-1/3} \qquad (2.91b)$$

for the unstable case ($z/L_\otimes < 0$), where air is colder than the ground. In Equations 2.91a and 2.91b, β is an empirical constant set to 35 W/m², $T_\otimes$ (in degrees Kelvin) is the turbulent temperature scale, z (in meters) is elevation, k (dimensionless) is the von Karman constant taken to be $k = 0.35$, and $L_\otimes$ (in meters) is the Monin-Obuhkov length.

Near the ground, the intensity of temperature fluctuations depends on the type of ground cover, soil humidity, ambient temperature, solar irradiation, and wind speed. Thiermann's model combines information on these environmental characteristics and provides solutions for C_T^2 that are expressed in terms of the friction velocity, $u_\otimes$ (in m/s); the turbulent temperature scale, $T_\otimes$ (in degrees Kelvin); and the Monin-Obuhkov length, $L_\otimes$ (in meters). These quantities can be expressed as [71]

$$u_\otimes = uk \left[\ln\left(\frac{z_u}{z_0}\right) - \psi(L_\otimes) \right]^{-1}, \tag{2.92a}$$

$$T_\otimes = -\frac{Q_0}{u_\otimes}, \tag{2.92b}$$

$$L_\otimes = \frac{u_\otimes^2 T}{kg T_\otimes} \tag{2.92c}$$

where u (in m/s) is the wind velocity measured at height z_u (in meters) above the ground, z_0 (in meters) is the roughness length of the ground surface, Q_0 (in degrees Kelvin × m/s) is the vertical turbulent kinematic heat flux, T (in degrees Kelvin) is the air temperature, and g is the acceleration of gravity (9.81 m/s²). Calculation of $u_\otimes$ depends on the form of $(\Psi L_\otimes)$, which is chosen according to whether the conditions are stable or unstable. The parameter $\Psi(L_\otimes)$ is given by [89]

$$\psi(L_\otimes) = \begin{cases} 2\ln\left(\frac{1+y}{2}\right) + \ln\left(\frac{1+y^2}{2}\right) - 2\tan^{-1} y + \frac{\pi}{2}, & z_u/L_\otimes < 0, \quad unstable \\ -4.7 z_u/L_\otimes, & z_u/L_\otimes > 0, \quad stable \end{cases}$$

$$\tag{2.93}$$

where $y = (1 - 15 z_u/L_\otimes)^{1/4}$. The kinematic heat flux is different during the daytime vs. nighttime [71, 90]:

$$Q_0 = \begin{cases} \dfrac{\eta}{c_p \rho}\left[\left(1 - \dfrac{\alpha_H}{1 + \gamma_H/s}\right)(1 - A)R - \beta\right], & daytime \\[4mm] \dfrac{cu^3}{1 + (c \cdot c_p \rho / H_{max})u^3}, & nighttime \end{cases} \tag{2.94}$$

where

$$c = -\frac{4}{27} \cdot \frac{k^2 T}{5 g z_u [\ln(z_u/z_0)]^2}.$$

In Equation 2.94, η (dimensionless) is an empirical constant equal to 0.9 [90], c_p (in J/kg/degrees Kelvin) is the specific heat of air, ρ

(in kg/m^3) is the density of air, and α_H (dimensionless) is a parameter between 0 (dry areas without vegetation, desert, rock land) and 1 (grass-covered land) that represents the capability of humidity in the ground to evaporate [90]. Parameter γ_H is the ratio of the specific heat of air at constant pressure to the latent heat of water vapor, and s is the temperature derivative of saturation-specific humidity. The quantity A (dimensionless) is the surface albedo and R (in W/m^2) is the solar irradiance. The heat flux H_{max} depends on local vegetation, ground characteristics, air humidity, and cloud cover and is between $H_{max} = -5$ W/m^3 and -100 W/m^3.

A multitude of physical quantities is required in the calculations. The specific heat of air, c_p, is taken to be a constant equal to 1004.6 J/kg/K [91], the density of dry air, ρ, and the ratio γ_H/s are given in the following formulas where temperature T is in degrees Celsius [90, 91]:

$$\rho = 1.286 - 0.00405T, \tag{2.95}$$

$$\gamma_H/s = 1.4631 - 0.0923T + 0.0027T^2 - 3.18 \cdot 10^{-5} T^{-3} \tag{2.96}$$

The Tiermann model was examined in experiments in Israel, in the Golan Heights mountains in the North and the Negev desert in the South.

The roughness length chosen was 0.02 m for both experimental locations. The ground humidity parameter α was set at 1 for the Golan and 0 for the Negev experiments (see previous section), respectively, because those locations were covered completely by grass or low bushes (Golan) and rocky desert (Negev). The value of ground albedo, A, used is 0.5 for Golan and 0.35 for the Negev experiments, based on measurements made with an albedometer and is for grass-covered and rocky surfaces [90].

For the Golan and Negev experiments, the measurements of C_n^2 and model calculations are shown in Figures 2.8 and 2.9, respectively. The irradiance, air temperature, and wind speed data were used as inputs to the model. Agreement between the measurements and model is generally good during the day, but not nearly so good at night.

The main reason the modeled turbulence parameters are in poorer agreement with the measurements at night is because the nighttime heat flux estimated by Equation 2.94 is not as accurate as that estimated for daytime. During the day, heat flux is driven by solar

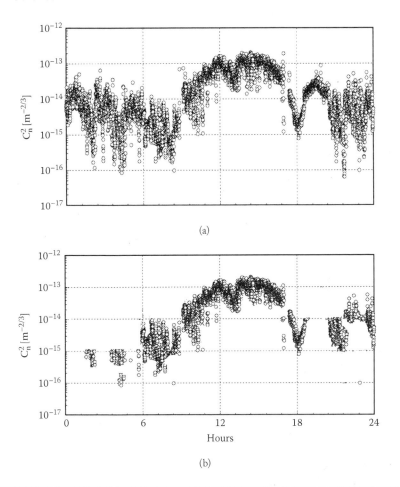

Figure 2.8   Golan coastal areas: plots of C_n^2 (a) measured and (b) modeled. (From S. N. Bendersky, N. Kopeika, and N. Blaunstein, "Atmospheric optical turbulence over land in middle east coastal environments: Prediction modeling and measurements," *Applied Optics*, vol. 43, pp. 4070–4079, 2004. With Permission.)

irradiation, a parameter that can be easily measured. At night there is no such dominant factor to which the heat flux can be attributed. Another limitation of this model is at wind speeds less than 1 m/s. In these cases, the model could not be used due to restrictions in the applicability of MOS (macroscale optical simple) theory under very stable conditions. Similar data and limitations have been obtained in investigations elsewhere [50].

2.7.2.2 Macroscale Meteorological Model   In contrast to the previous models, the prediction of turbulence values by macroscale meteorological

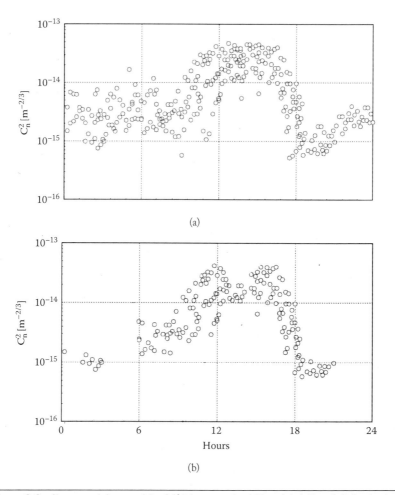

Figure 2.9 Negev coastal areas: plots of C_n^2 (a) measured and (b) modeled. (From S. N. Bendersky, N. Kopeika, and N. Blaunstein, "Atmospheric optical turbulence over land in middle east coastal environments: Prediction modeling and measurements," *Applied Optics*, vol. 43, pp. 4070–4079, 2004. With Permission.)

models [32, 46] is carried out without a long calculation algorithm and several additional pieces of data (irradiance and heat flux, for example). This model is based on the concept of temporal hours or relative part of the day and is a fairly reliable way to predict C_n^2. In units of $m^{-2/3}$,

$$C_n^2 = 3.8 \cdot 10^{-14} W + f(T) + f(U) + f(RH) - 5.3 \cdot 10^{-13} \quad (2.97)$$

where

$$f(T) = 2 \cdot 10^{-15} T \tag{2.98a}$$

$$f(U) = -2.5 \cdot 10^{-15} U + 1.2 \cdot 10^{-15} U^2 - 8.5 \cdot 10^{-17} U^3 \tag{2.98b}$$

$$f(RH) = -2.8 \cdot 10^{-15} RH + 2.9 \cdot 10^{-17} RH^2 - 1.1 \cdot 10^{-19} RH^3 \tag{2.98c}$$

Here, W is temporal hour weight, well-recognized and described in References 29 and 72; T is air temperature (in degrees Kelvin); U is wind speed (in m/s); and RH is relative humidity (in %). This model for C_n^2 applies to about 15 m of elevation. Dynamic range for temperature is from 9 to 35°C, for relative humidity from 14 to 92%, and for wind speed from 0 to 10 m/sec.

For elevations other than 15 m (e.g., 2.5 m in our experiments), the previously calculated value of the refractive index structure parameter can be scaled according to various models of the height profile of C_n^2. Although many models have been suggested, experiments described in Reference 92 for measurements up to 100 m elevation support primarily the model of Tatarskii [23], which gives

$$C_n^2(h) = C_{n0}^2 h^{-4/3} \tag{2.99}$$

where C_{n0}^2 is the refractive index structure coefficient at the surface. This height profile would no longer be valid at the boundary level where C_n^2 suddenly decreases rapidly as elevation increases.

Equations 2.97 and 2.98 contain four kinds of regression coefficients (i.e., four terms). The first term in Equation 2.97 is a coefficient of the temporal hour weight function that contains information about solar radiation by the hour between sunrise and sunset (must be positive [72]). The second term in Equation 2.97 is the temperature coefficient (defined by Equation 2.98a) and undoubtedly should be positive because a higher temperature usually leads to a larger temperature gradient and hence to stronger turbulence. The third term characterizes the wind speed coefficients (determined by Equation 2.98b). They are expected to be negative because wind causes air mixing and therefore decreases the inhomogeneity of temperature and humidity and hence, according to References 77 and 93, decreases C_n^2.

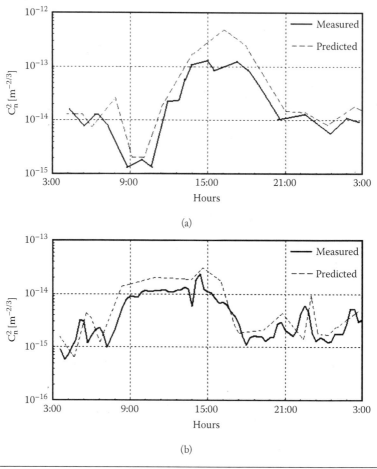

Figure 2.10 Comparison between measured and "practical" modeled average C_n^2 values: (a) Golan and (b) Negev experiments. (From S. N. Bendersky, N. Kopeika, and N. Blaunstein, "Atmospheric optical turbulence over land in middle east coastal environments: Prediction modeling and measurements," *Applied Optics*, vol. 43, pp. 4070–4079, 2004. With Permission.)

Also, as the wind increases, so too does dissipation of ground heating, thereby also decreasing temperature gradient and C_n^2. The last term in Equation 2.97 determines the relative humidity coefficients. High relative humidity is usually related to low-temperature and low-humidity gradients [77, 93], so negative values are expected and are indeed obtained (according to Equation 2.98c). The model described by Equation 2.97 has been validated over both desert surfaces and high-density vegetation surfaces.

Figure 2.10a and b presents a comparison between measured and predicted magnitudes of the refractive index parameter C_n^2 via

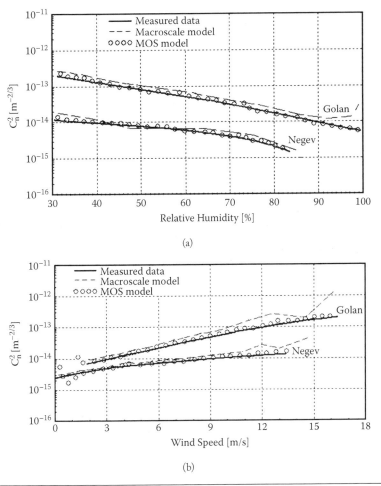

Figure 2.11   Comparison between measured and modeled C_n^2 values versus (a) relative humidity and (b) wind speed data. (From S. N. Bendersky, N. Kopeika, and N. Blaunstein, "Atmospheric optical turbulence over land in middle east coastal environments: Prediction modeling and measurements," *Applied Optics*, vol. 43, pp. 4070–4079, 2004. With Permission.)

this macroscale model for the Golan and Negev experimental data, respectively. It is shown that the macroscale meteorological approach can be used successfully to predict C_n^2.

The limitations of these models, "practical" macroscale [72] and theoretical [87] based on Thiermann's MOS theory application [71], follow from results shown in Figure 2.11a and b, where C_n^2 is described versus relative humidity and wind speed, respectively, for both experimental locations.

The practical estimation model of C_n^2 showed high correlation (up to 90%) with measured results over a wide range of meteorological conditions, but not including conditions involving large values of relative humidity (>92%) and high wind speeds (>11 m/s). For these limited cases the MOS theory is more reliable for prediction of C_n^2 values (see Figure 2.7). However, on the other hand, for wind speeds less than 1 m/s and nighttime, Thiermann's model could not be used due to restrictions in the applicability of MOS theory under very stable conditions. Thus, in spite of the limitations of both models, they supplement each other for prediction of C_n^2.

2.7.2.3 Extension of the Macroscale Meteorological Model The macroscale meteorological model predicts very accurate results both for nighttime and for daytime measurements, in which data such as soil type and vegetation affect the temperature gradients and humidity measurements that are already reflected in the model. However, in the previous section we showed the limitation of this model for prediction of C_n^2 over midland coastal zones, when large values of relative humidity (>92%) and high wind speeds (>11 m/s) are involved.

Strong wind has a positive effect on turbulence by reducing it. The wind causes the atmosphere's index of refraction to be homogeneous and thus reduces the gradient. However, winds stronger than about 15 knots (>7.7 m/s) do not reduce air circulation but rather increase turbulence, in addition to increasing the number of particles in the air (such as dust) and, as a result, increase the radiance attenuation and scattering. They also contribute to absorption of radiation by aerosols and thus to increased atmospheric heating and, therefore, to increased C_n^2 [77–83].

Generally speaking, as the relative humidity increases, the degradation of the image resolution due to the influence of turbulence is noted. The structure parameter of the refractive index C_n^2 appears in many formulae that characterize optical turbulence. Basic formulae of C_n^2 include the gradient of the actual refractive index as a coefficient, which is a direct function of the temperature and relative humidity gradients. Changes in humidity, especially in low-humidity conditions, cause a random change in the refractive index. The two factors, temperature and humidity, vary more in proximity to the ground.

The following versions of the macroscale practical meteorological model, presented for daytime and nighttime separately, may extend this model, and include influences of both high wind speed and high relative humidity, as well as the type of land (albedo values, A) and the temporal time hours. Experimentally, it was found that this extended model is valid for 9 to 35°C temperature and a height of 2.5 m above ground.

The following equation shows prediction of C_n^2 over land for mid-land coastal zones, during daytime, and is valid for wind speed values of 8 m/s $\leq U \leq 17$ m/s and relative humidity of $30\% \leq RH \leq 70\%$ [78]:

$$C_n^2 = 3.8 \cdot 10^{-14} W + \frac{A}{\exp[T]} \cdot 10^{-4} + f(U) + f(RH) - 4.45 \cdot 10^{-14}$$

(2.100)

where, for *rocky land* (*rocky desert*), the following functions on statistical basis are

$$f(U) = 8 \cdot 10^{-16} U - 4 \cdot 10^{-18} U^2 \qquad (2.101a)$$

$$f(RH) = -8 \cdot 10^{-16} RH + 5 \cdot 10^{-18} RH^2 \qquad (2.101b)$$

and, for *vegetation-covered land*,

$$f(U) = 2.58 \cdot 10^{-14} U \qquad (2.102a)$$

$$f(RH) = -6.797 \cdot 10^{-15} RH \qquad (2.102b)$$

where W is the temporal hour weight [72, 78] and A is the value of ground albedo. For the Golan and Negev experiments, the values of A used were 0.5 and 0.35, respectively, based on measurements made with an albedometer. These are typical values for vegetation-covered and rocky lands [90]. T is air temperature (in degrees Celsius), U is wind speed (in m/s), and RH is relative humidity (in %).

During nighttime, prediction of the turbulence parameter over land is shown by Equation 2.102, which is valid for wind speed 5 m/s $\leq U \leq 10$ m/s and for high values of relative humidity, $92\% \leq RH \leq 100\%$. This formula is acceptable just for the vegetation-covered midland coastal region (Golan coastal zone), because during the Negev experiment a data fit for such cases of relative humidity and wind speed parameters was not obtained. Instead, it was found that [78]

$$C_n^2 = f(T) + f(U) + f(RH) - 1.9 \cdot 10^{-14} \qquad (2.103)$$

where

$$f(T) = 3 \cdot 10^{-17} T \qquad (2.104a)$$

$$f(U) = 1.2 \cdot 10^{-14} U \qquad (2.104b)$$

$$f(RH) = -7.5 \cdot 10^{-16} RH \qquad (2.104c)$$

This last sub-model extends the meteorological practical model without using the temporal hour and ground albedo parameters, because at nighttime, for high relative humidity and wind speed values, these parameters do not play a major role. The graphical comparison between measured and modeled data, based on practical meteorological sub-models (Equations 2.100 and 2.103), is presented in Figure 2.12. Figure 2.12a shows the normalized measured and prediction data of C_n^2 versus high relative humidity values. In this case, the limitation of the original macroscale practical model is clearly seen. The extended new sub-model predicts the turbulence parameter with high correlation (up to 90%) with measured results over a selected range of meteorological conditions. The same good correlation between the measured and theoretical prediction data of C_n^2 variations for high wind speed values is clearly seen from the results presented in Figure 2.12b.

We can stress that the extended model presented here with two variants for daytime and nighttime over the Negev coastal zone atmosphere yields excellent results for the specific coastal experimental sites. Furthermore, both proposed formulas (2.100 and 2.103) are mathematically simple for further computations for various meteorological situations. In addition, these sub-models were tested by use of the specific characteristics of the experimental sites, from which information about the albedo parameter was obtained.

These results support the experimentally proved simple meteorological model (with two variants for daytime and nighttime periods) for practical estimations of the refractive index structure parameter, C_n^2, in the turbulent atmosphere over land in coastal zones. This simple model is a macroscale meteorological model [72, 73, 78] that is

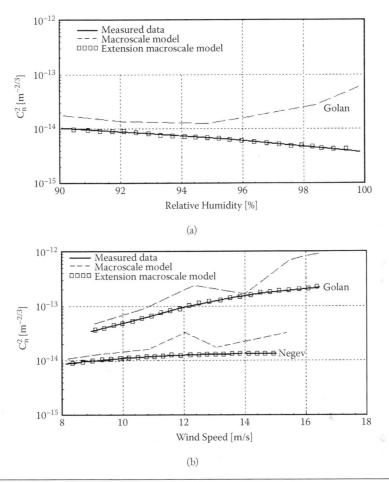

Figure 2.12   Comparison between measured and modeled C_n^2 values versus: (a) relative humidity data for Golan coastal zone, normalized to $U = 8$ m/s and $T = 20°C$, and (b) wind speed data for both Negev and Golan coastal zone, normalized to $RH = 40\%$ and $T = 20°C$. (From S. N. Bendersky, N. Kopeika, and N. Blaunstein, "Atmospheric optical turbulence over land in middle east coastal environments: Prediction modeling and measurements," *Applied Optics*, vol. 43, pp. 4070–4079, 2004. With Permission.)

extended for high values of wind speed and relative humidity, as well as the model developed in Reference 87 according to Thiermann's concept [71] based on MOS theory [70].

Values of the C_n^2 parameter calculated by all three models were compared with measurement data. It was shown that the macroscale meteorological approach [72, 78] can be used successfully to predict values of atmospheric turbulence. This practical model of C_n^2 variations shows high correlation (up to 90%) with measured results over a wide range of meteorological conditions, except when large values of relative humidity (>92%) and high wind speeds (>11 m/s) occur.

This practical meteorological model, which covers both daytime and nighttime conditions and is based on experimental data, extends the existing macroscale model for these specific weather conditions as a good predictor (with high degree of correlation, up to 90%) of the refractive structure of over-land atmospheric turbulence. In addition, it was found that predictions of Thiermann's model are in good agreement with the measured values during unstable daytime conditions but have poor agreement with the nighttime experimental data. Also, for wind speeds less than 1 m/s, which often occurred at night, the Thiermann's model cannot be used due to restrictions in the applicability of MOS theory under very stable conditions.

In spite of these limitations, values of C_n^2 calculated with Thiermann's model for daytime conditions and with the practical meteorological models for both night and daytime conditions are accurate enough to provide useful estimates of optical turbulence effects on electro-optical or laser systems over land coastal zones.

The validity of all models was examined in different geographical areas with different surface cover, varying from a completely desert location in the Eilat (Negev) coastal zone to an intensely vegetated surface in the Tiberias (Golan) coastal zone [77–79]. The measurements have shown very good agreement with prediction by the new practical meteorological model, both for daytime (according to Equation 2.100) and nighttime (according to Equation 2.103). Knowledge of height profiles for the C_n^2 parameter in the boundary layer by Tatarskii [23, 69] can be used to translate this prediction to other elevations.

At the same time, we should stress that the extension of the practical macroscale meteorological model can be quite relevant only for horizontal atmospheric optical paths. Despite this fact, and that the experiments were carried out for only two midland coastal zones, we suggest that the proposed model, as an extension of the macroscale meteorological model, can be successfully used for the same coastal zones as well as for typical marine areas. It broadens the models introduced in References 72 and 78 and in References 70, 71, and 87 for marine (near-the-sea) and land coastal atmospheric environments, respectively. This can be verified only by experiments in other areas.

2.8 Line-of-Sight Bending Caused by Strong Atmospheric Turbulence

Line-of-sight bending, called "beam bending," is an effect caused by various meteorological phenomena, including strong atmospheric turbulence, occurring in the non-regular layered atmosphere. During daytime the desired targets appear to be lower and during nighttime higher than their actual locations. Bending is more pertinent for low wind speed and high turbulence conditions.

Over the past decade, predictions of laser beam distortions between sensor and target caused by micro and macro weather atmospheric effects have been shown to be important to optical systems of various applications [29, 94–98]. Differential heating [71, 73, 99, 100] of desert and vegetation regions, valley and mountain areas, and buildings and other structures results in thermal clutter that can confuse infrared sensors [32, 40]. Additionally, the radiative transfer and optical properties of the atmosphere can be affected by scattering and absorption [29, 72, 101, 102] of aerosol particles. Other parameters that must be considered include the contrast between the target and its background, the path radiance, refraction, turbulence, and molecular effects [29, 103–105]. Changes in molecular density and the resulting gradient in the index of refraction cause bending of a light ray traveling through the atmosphere [66, 67]. For example, over long paths at sunset, refraction causes such phenomena as flattening of the solar disc and the "green flash." Strong temperature gradients over short paths are responsible for such effects as mirages and looming [107–109].

Atmospheric refraction has dramatic effects on the performance of target detection [32, 40, 60], particularly for near-ground long distances (2 km and more) between targets and observation points. Ducting can result in greatly extended ranges, whereas sub-refracting conditions may impose severe limitations [110, 111]. Sets of models, each with their advantages and drawbacks, are available to predict the likely effects of refraction on an optical beam, but the weather often changes rapidly, and it is difficult to model its effects accurately [93, 112, 113].

Line-of-sight bending of optical waves between target and sensor, based on atmospheric refraction, is mainly affected by strong atmospheric turbulence. The turbulent atmosphere causes the intensity of a wave beam to fluctuate or scintillate and gives rise to beam wander and distortion and random displacement of images [29, 77–80, 86, 88].

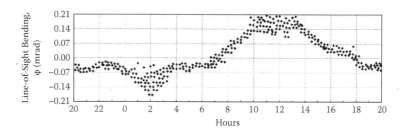

Figure 2.13 Typical whole-day measurements of line-of-sight bending for 2.5 m height level in Golan experimental areas (May 2002). φ is positive for downward bending.

In this section, we present theoretical derivation and a semi-empirical model for line-of-sight bending prediction based on turbulence and atmospheric condition parameters such as average atmospheric pressure, temperature, relative humidity, and average wind speed values. Then we analyze, following Reference 79, the line-of-sight bending values and their dependence on refractive index structure parameters (C_n^2), meteorological parameters such as wind speed, and source and detector elevation.

In general, beam bending over land is downward in daytime, upward at nighttime, and approaches zero shortly before sunset and shortly after sunrise, corresponding to the diurnal minima of C_n^2 [79].

Figure 2.13, extracted from Reference 79, shows several typical measured data of line-of-sight bending made at a height of 2.5 m during 24 hours in the Golan area. The symbol near line-of-sight bending values on the vertical axis describes the atmospheric physical conditions: stable (negative values) or unstable (positive values), where the refractive index decreases or increases with height profile, respectively.

Thus, based on maximal values of beam bending (0.21 mrad) for the 2.5 m height Golan measurements over a 2470 m propagation path, we can expect the far target to be about 0.52 m lower (daytime) or higher (nighttime) relative to real height above the ground. Figure 2.14 shows the Negev experiments' (September 2003) daytime 2.5 m height measurements of line-of-sight bending versus wind speed values. Here, the far target over the 3760 m path showed 0.25 mrad daytime bending of 0.95 m downward relative to actual height [79].

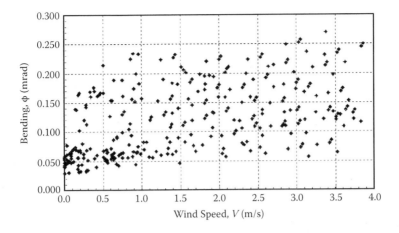

Figure 2.14   Plot of line-of-sight bending versus wind speed measurements for 2.5 m height in the Negev area (September 2003).

2.8.1 Modeling of Line-of-Sight Bending

Beam bending over near-ground long atmospheric optical paths is caused by variations (atmospheric turbulence) of the refractive index n of the atmosphere along the optical propagation channel, which is a function of the various meteorological parameters [114–116]. For example, the refractive index (ignoring minor wavelength dependence) for a marine atmosphere may be predicted by [29, 79, 117]

$$n \approx 1 + N = 1 + 77.6\frac{P}{T} \cdot \left(1 - 7733\frac{q}{T}\right) \cdot 10^{-6}, \qquad (2.105)$$

where P is air pressure (millibars), T is temperature (degrees Kelvin), and q is specific humidity (g/m³). By introducing in Equation 2.105 the typical values for pressure, temperature, and humidity, one can see that $N \ll 1$.

Nevertheless, the fluctuations in N, which depend largely on temperature and humidity gradients, strongly affect the image quality and distortion because of the large number of random refractions that a light beam undergoes while propagating a relatively long distance through the atmosphere. The randomness in both time and space domains causes the incident light to be received at a large variety of angles of incidence [23, 29, 69, 79, 114], because the ground is generally warmer than the surrounding air during the day and cooler at night.

Temperature fluctuations arise because of the moving heated air near the ground and the existence of a thermal gradient in the atmosphere that is usually greatest near the ground. Wavefront tilt described by Equation 2.91 arises from random changes in air temperature and pressure, which lead to random changes in N. Humidity fluctuations, particularly at low relative humidity [93], give rise to similar effects. Such random refractions cause random wavefront tilt and bending that, when integrated over many tilt angles during an exposure, give rise to image blur and distortion caused by optical turbulence.

At the same time, as was shown in References 23, 29, 69, 79, and 114, the turbulence appearing near the ground is strongly dependent on the height from the surface and the surface conditions. In addition, the turbulence is affected by change in the average wind velocity with height [70, 117–119]. We show briefly both physical and empirical frameworks accounting for meteorological conditions and near-ground turbulence, stemming from fluctuations of the refractive index. These predictions were compared with measured data, and an extended new model that accounts best for day and night predictions is presented here.

2.8.2 Boundary Layer Turbulence Modeling

Temperature and water vapor fluctuations are known to be the primary mechanisms mediating the effects of atmospheric turbulence on the index of refraction [23, 29, 69, 114–117], where pressure fluctuations can be neglected in the real atmosphere. Recent studies of correlation coefficient at microwave frequencies, where the refractive index is much more sensitive to humidity than at optical and visual frequencies, indicated, however, that wind speed was the dominant meteorological parameter affecting turbulence. When wind speed is too high (about 8 m/s), it decreases air mixing and therefore increases turbulence strength.

For turbulence in the boundary layer, an important parameter is the change in the average horizontal wind velocity with height [21, 92]. For the few hundred meters near the ground, the wind velocity is known to have a logarithmic profile with height; whereas on the ground, owing to viscosity, wind velocity is zero [21, 120]:

$$V(z) = (V_*/0.4)\ln(z/z_0) \qquad (2.106)$$

where V_* is the friction velocity (constant), z is the height above the ground surface, and z_0 is the roughness length. The roughness length should depend only on surface roughness. Typical roughness lengths over uniform terrain, for example, equal 0.01 m for mown grass, 0.05 m for long grass or rocky ground, 0.2 m for pasture land, 0.6 m for suburban housing, and from 1 to 5 m for forests or urban areas.

The turbulence is characterized by the energy dissipation rate, ε, where the thermodynamic process of turbulence is adiabatic [120]. The energy dissipation must be balanced by the sum of the wind shear energy, M, and the buoyancy energy term, B. The rate of production of turbulent energy by wind shear [121, 122], M, and rate of production of energy by buoyancy [121–123], B, are

$$M = K_m \left(\frac{\partial V}{\partial z} \right)^2, \quad B = -K_b \frac{g}{\theta} \frac{\partial \theta}{\partial z} \qquad (2.107)$$

where K_m and K_b are coefficients of "eddy viscosity" and "eddy coefficient of heat conduction" (as was defined in References 120–123), respectively; g is gravitational acceleration [9.81 J/(kg · m), for standard atmosphere]; and $\theta = T + \alpha_a z$ is potential temperature [123], where T is atmospheric temperature and $\alpha_a = 0.98$ °K/100 m is called the adiabatic rate of the decrease in temperature. The atmosphere is considered to be "neutral" when the potential temperature of the atmosphere is the same at all heights [120–123]. The atmosphere is said to be "stable" if the potential temperature increases with height (T decreases at a rate less than α_a), so the air, if displaced up or down, becomes cooler or warmer than the surrounding air and thus returns to its original position [21, 120–123]. If the potential temperature decreases with height, the air will not return to its original position and thus the atmosphere is "unstable." Therefore, at nighttime the atmosphere is stable ($\partial \theta / \partial z > 0$); whereas in daytime the atmosphere is unstable ($\partial \theta / \partial z < 0$). The ratio of the shear energy production, M, to the buoyancy, B, is connected to the Richardson number, R_i [82–84]:

$$R_i = \frac{(g/\theta)(\partial \theta / \partial z)}{(\partial V / \partial z)^2} \qquad (2.108)$$

Using Equation 2.108, we obtain, according to Reference 21, $\partial V / \partial z = V_* / 0.4z$. However, from experimental data a small correction is needed [21, 124]:

$$\left(\frac{\partial V}{\partial z}\right) = \left(\frac{V_*}{0.4z}\right)\phi, \tag{2.109}$$

where ϕ is unity for neutral atmosphere [21, 81, 82]. The eddy viscosity coefficient $K_m = 0.4V_*z/\phi$. Thus, the energy dissipation rate is [21, 124]

$$\varepsilon = M + B = (V_*^3/0.4z)\phi_\varepsilon, \tag{2.110}$$

where ϕ_ε includes ϕ as well as the effect of B. In terms of the Richardson number, the empirical expressions for unstable [21, 122] air conditions are

$$\phi_\varepsilon = (1 - 18R_i)^{-1/4} - R_i, \quad \phi = (1 - 18R_i)^{-1/4}, \tag{2.111a}$$

and for stable [21, 82] air conditions are

$$\phi_\varepsilon = (1 - 0.7R_i)/(1 - 7R_i), \quad \phi = R_i/(1 - 7R_i). \tag{2.111b}$$

Finally, the temperature structure constant, C_θ^2, and the structure constant of the index of refraction, C_n^2, can be calculated in the following forms [21, 121–123]:

$$C_\theta^2 = b(0.4z)^{4/3}\phi^{-2}\phi_\varepsilon^{-1/3}(\partial\langle\theta\rangle/\partial z)^2,$$
$$C_n^2 = b(0.4z)^{4/3}\phi^{-2}\phi_\varepsilon^{-1/3}(\partial\langle n\rangle/\partial z)^2, \tag{2.112}$$

where b is considered to be a universal constant estimated to be in the range 1.5–3.5; $b = 2.8$ is suggested by References 21 and 125. Therefore, it can be seen that the temperature structure and the index of refraction structure parameters are dependent on the gradient of the mean potential temperature and on the mean index of refraction, respectively. For optical wave propagation, the structure constant C_n^2 of the index of refraction may be written in the form of Equation 2.90, which we will present here again for the further convenience of current discussions:

$$C_n^2 \cong \left(79 \cdot 10^{-6}\frac{P}{T^2}\right)^2 C_\theta^2, \tag{2.113}$$

where, for stable and unstable atmospheric conditions, we may use the more comprehensive term for potential temperature structure constant [119, 126]:

$$C_\theta^2 = z^{4/3}(\partial\theta/\partial z)^2 f_3(R_i). \tag{2.114}$$

Here the Richardson number, R_i, varies from -2.0 to 0 for unstable (daytime) and from 0 to 0.20 for stable (nighttime) atmospheric conditions [126]. The function $f_3(R_i)$ varies from 3.62 to 1.09 for unstable (daytime) and from 1.09 to 0.015 for stable (nighttime) atmospheric conditions [126]. The gradient of potential temperature is well known to be of the form [21, 126]

$$\frac{\partial\theta}{\partial z} = \frac{1}{z}\left[\frac{T_1 - T_2 + \alpha_a(z_1 - z_2)}{\ln(z_1/z_2)}\right], \tag{2.115}$$

where T_1 and T_2 are measured temperatures at two different heights, z_1 and z_2, and $\alpha_a = 0.0098$ K/m. Substituting this expression in Equation 2.114, we can evaluate C_n^2 from Equation 2.113.

Many other models have been developed for refractive index structure parameter C_n^2 prediction, but all of them functionally use the standard meteorological parameters as input [23, 29, 69–72, 77, 86, 88, 99, 117–119]. However, following a simple empirical model [29, 72], obtained for different Israeli experimental areas, we can predict C_n^2 in the more general form following results obtained in Reference 78, which are summarized by Equations 2.103 and 2.104. Here, different functions of air temperature, wind speed, and relative humidity are considered versus experimental areas: rocky and desert land, vegetation-covered land, continental areas, and coastal areas. The parameter of *const* is a regression coefficient dependent on experimental definitions. The dynamic range of this model for temperature is from 9 to 35°C, for humidity from 14 to 100% [72, 78], and for wind speed from 0 to 17 m/s [29, 78, 79]. These models for C_n^2 have been supported by measurements in France, India, and the United States under these weather conditions. Equation 2.113, with additional functions and parameters defined by Equations 2.114 and 2.115, was based on measurements in 1990 at Beer-Sheva, validated using measurements in 1992–1993 at Sede-Boqer [72, 86], and extended using measurements in 2001–2003 in the Negev and Golan experiments [78, 79].

The meteorological data must be measured at similar heights according to the demands of this model. For other elevations, the previously calculated value of the refractive index parameter on chosen height can be scaled according to various models of the height profile of C_n^2.

2.8.3 Fluctuations of the Refractive Index

The variations of the index of refraction with potential temperature θ and with the specific humidity q are directly related to the fluctuation characteristics of the index of refraction in turbulence media [21, 23, 69]:

$$\frac{\partial n}{\partial \theta} = \frac{77.6P}{(\gamma-1)T^2}\left(1 + \frac{15466q}{T}\right)\cdot 10^{-6}, \quad \frac{\partial n}{\partial q} = \frac{77.6 \cdot 7733}{T^2}\cdot 10^{-6}$$

(2.116)

where $\gamma = C_p/C_v$ is the ratio of the specific heat at constant pressure to constant volume. For the standard atmosphere, $C_p = 0.240$ (cal/°C) and $C_v = 0.171$ (cal/°C).

The gradient of the mean index of refraction is given by [21, 23, 69]

$$\frac{\partial \langle n \rangle}{\partial z} = \frac{\partial n}{\partial \theta}\frac{\partial \theta}{\partial z} + \frac{\partial n}{\partial q}\frac{\partial q}{\partial z}.$$

(2.117)

The humidity variation with height (gradient of humidity) has a negligible effect on optical and infrared wave horizontal propagation [21, 29, 114]. Therefore, $\partial\langle n\rangle/\partial z \approx (\partial n/\partial\theta)(\partial\theta/\partial z)$, where the $(\partial\theta/\partial z)$ term may be determined from Equation 2.115. On the other hand, it can be calculated using models, described by Equations 2.113 and 2.114.

Thus, it can be seen that all models adequate for $(\partial n/\partial\theta)$ and $(\partial\theta/\partial z)$ representation, or direct models for index of refraction fluctuations $(\partial n/\partial z)$, can be relevant to atmospheric turbulence or line-of-sight bending predictions. For such media, especially for inhomogeneous media, a beam does not travel in a straight line because of continual changes in the refractive index (see Figure 2.15).

The *ray equation* is used to describe changes in the refractive index as a function of location, r, and path length, s, of a ray propagation [21, 29]:

$$\frac{d}{ds}\left(n\frac{dr}{ds}\right) = \nabla n$$

(2.118)

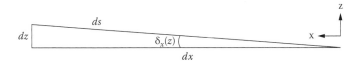

Figure 2.15 Schematic plot of ray propagation nearly parallel to an optical axis in inhomogeneous media.

where, for a homogeneous medium, $\nabla n = 0$ and light rays take straight-line paths. For the paraxial rays in inhomogeneous media in Figure 2.15, Equation 2.118 can be rewritten in the form

$$\frac{d}{ds}\left(n\frac{dz}{ds}\right) = \frac{\partial n}{\partial z} ,$$

(2.119a)

or

$$\frac{d}{ds}\left(n\frac{dx}{ds}\right) = 0,$$

(2.119b)

where $(dz/ds) = \sin[\delta_x(z)]$ and $(dz/ds) = \cos[\delta_x(z)]$. From Equation 2.119b, we get:

$$n(z)\cos[\delta_x(z)] = const = \beta$$

(2.120)

Therefore,

$$\cos[\delta_x(z)] = \beta/n(z)$$

(2.121)

and

$$\tan[\delta_x(z)] = \frac{dz}{dx} = \frac{\sqrt{n^2(z) - \beta^2}}{\beta}$$

(2.122)

Because $\cos[\delta_x(z)] = (dx/ds) = [\beta/n(z)]$, then

$$ds = \frac{n(z)dx}{\beta}.$$

(2.123)

This is substituted into Equation 2.119a to obtain

$$\frac{d}{dx}\frac{\beta}{n(z)}\left[n(z)\frac{dz}{dx}\frac{\beta}{n(z)}\right] = \frac{\partial n}{\partial z}$$

(2.124)

Although refractive index n varies with elevation z, we assume it does not vary along horizontal path x. Therefore, for a given height z, $n(z)$ is constant. Accordingly, Equation 2.124 reduces to

$$\beta^2 \frac{d^2 z}{dx^2} = n(z) \frac{\partial n(z)}{\partial z},$$
(2.125)

or, from Equation 2.121,

$$\frac{d^2 z}{dx^2} = \frac{1}{\beta^2} n(z) \frac{\partial n}{\partial z} = \frac{1}{\cos^2[\delta_x(z)]} \frac{1}{n(z)} \frac{\partial n}{\partial z}$$
(2.126)

Alternatively, Equation 2.126 can also be obtained by differentiating Equation 2.122 and using appropriate substitutions. Assuming further that the beam bending angle $\delta_x(z)$ in Equation 2.126 is small and that $n(z)$ is not only constant for given z but changes with height z much more slowly than does $\partial n(z)/\partial z$, then

$$\frac{d^2 z}{dx^2} \approx \frac{\partial n}{\partial z}$$
(2.127)

By integrating this expression twice according to dx, we obtain

$$z = \frac{x^2}{2} \frac{\partial n}{\partial z} + C_1 x + C_2$$
(2.128)

where C_2 is a constant equal to the initial height, z_0, and C_1 is the initial beam bending angle in the camera plane, which is a constant, although the beam bending increases in magnitude as the beam propagates in direction x toward the target. Because $C_1 x + C_2$ is therefore constant, we arrive at the final result:

$$z = \frac{x^2}{2} \frac{\partial n}{\partial z} + const,$$
(2.129)

from which the line-of-sight bending, φ, includes the term

$$\varphi \approx \frac{z}{x} = \frac{L}{2} \frac{\partial n}{\partial z}$$
(2.130)

where L is the propagation path length, equal in our experiments to 2470 and 3760 m. Other researchers suggest similar expressions for atmospheric ray tracing for horizontal near-surface propagation [127, 128].

2.8.4 Line-of-Sight Bending Prediction

Low wind speed and high turbulence strength conditions (a result of high temperature and humidity fluctuations) give rise to strong line-of-sight bending near ground levels. Beam bending is negligible and exhibited a change of direction from downward (daytime) to upward (nighttime) relative to the real location of the far target after sunrise and before sunset, when ground and surrounding air temperatures are closest.

Figure 2.16a and b, extracted from Reference 79, shows typical scatterplots of the beam bending values versus atmospheric turbulence

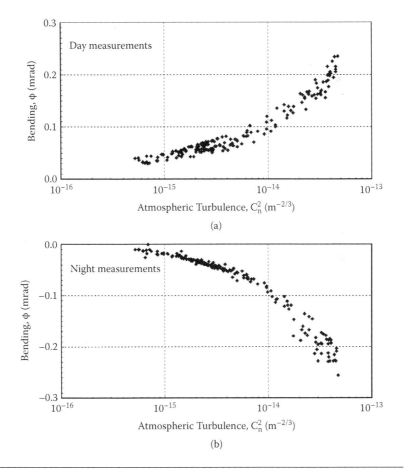

Figure 2.16 Plot of line-of-sight bending values versus refractive index structure parameters measured at 2.5 m above the ground in the Negev area (October 2001): (a) daytime and (b) nighttime data. φ is positive for downward bending. (From S. Bendersky, N. Kopeika, and N. Blaunstein, "Prediction and modeling of line-of-sight bending near ground level for long atmospheric paths," Proc. of SPIE International Conference, San Diego, CA, August 3–8, 2004, pp. 512–522. With Permission.)

values measured at 2.5 m above the ground. For other levels of height the relationship between φ and C_n^2 exhibits similar shapes or plots. For example, for 7.5 m heights, when wind speed is 1.5 m/s and $C_n^2 \cong 1 \cdot 10^{-13}$ m$^{-2/3}$, the bending maximal value decreased to 0.025 mrad, whereas for 10 m heights, it decreased to 0.005 mrad for similar meteorological conditions.

From Figures 2.13, 2.14, and 2.16, the change of direction fit can be seen for changing values of the refractive index structure parameter, C_n^2, suitable to transition between stable (where air is warmer than the ground) and unstable conditions [23, 29, 69, 78, 99, 121–123].

Unfortunately, for 2 m heights and 1.5 m/s wind speed, the bending maximal value is equal to 0.32 mrad, whereas for 4.5 m/s wind speed, it equals 0.38 mrad. It can be seen from Figure 2.16 that wind speeds have some role in bending effects, especially for low wind speed values.

Thus, as can be seen in the previous section, the vertical bending of the optical path of a beam of light propagating horizontally is caused by a vertical refractive index gradient. Several researchers [127, 128] have suggested simple models for prediction of the gradient of the refractive index; for example:

$$\frac{\partial n}{\partial z} = -78.2 \cdot 10^{-6} \frac{P}{T^2} \left(\frac{\partial T}{\partial z} + 0.0344 \right) \quad (2.131)$$

where the temperature gradient is obtained through application of similarity theory [21, 23, 69–71]. In general, the temperature gradient varies as z^{-1} for neutral conditions, as $z^{-4/3}$ for day conditions, and as $z^{-2/3}$ for night conditions. Through Equation 2.130, the semi-empirical Tofsted-Gillespie model [128] reasonably predicts the daytime bending, but it is not accurate for night conditions with large temperature gradients. The validity of this model was examined in Reference 88, and agreement with daytime measurements in References 72, 78, 79 was obtained when line-of-sight was less than 5 m above the terrain, according to the requirements of the model. This model predicts that the greatest temperature gradients occur when winds are low, skies are clear, and the soil is dry [88]. These conditions occur most often in deserts.

Figure 2.17 shows a comparison between the measured data and that predicted by the 2.130–2.131 model of line-of-sight bending for

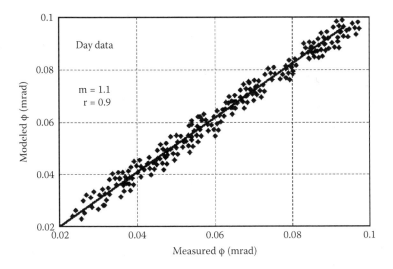

Figure 2.17 Scatterplots of the Tofsted-Gillespie model [88] line-of-sight bending versus measurements for 4 m height above the ground in the Negev experimental area during the day; *m* is the slope of regression line and *r* is the Pearson error value. (From S. Bendersky, N. Kopeika, and N. Blaunstein, "Prediction and modeling of line-of-sight bending near ground level for long atmospheric paths," Proc. of SPIE International Conference, San Diego, CA, August 3–8, 2004, pp. 512–522. With Permission.)

the Negev 4 m height daytime experiments. Therefore, for daytime in arid environments, that model can be useful at low elevations. However, a model for other conditions is desired.

Based on other well-known models for relation between the index refractive structure parameter, C_n^2, and gradients of wind speed, potential temperature, and refractive index (for example, the Panofsky-Dutton [121] and Wyngaard-Izumu-Collins [126] models presented by Equations 2.129 and 2.130–2.131, respectively), and by use of Equation 2.130, we can predict optical wave bending (e.g., line-of-sight bending).

Both models are available for unstable and stable (daytime and nighttime) conditions, and are found to be compatible with about 90% correlation for beam bending prediction in different desert and low-vegetation Middle East (Israel) areas (see Figure 2.18). But for stable (nighttime) conditions, where the ϕ and ϕ_ε functions used in the Panofsky-Dutton model [121] go to infinity, and in the cases where the Richardson number approaches 0.14, this model does not work well. Therefore, the semi-empirical Wyngaard-Izumu-Collins model [126] is more suitable for near-ground horizontal propagation of optical wave bending prediction.

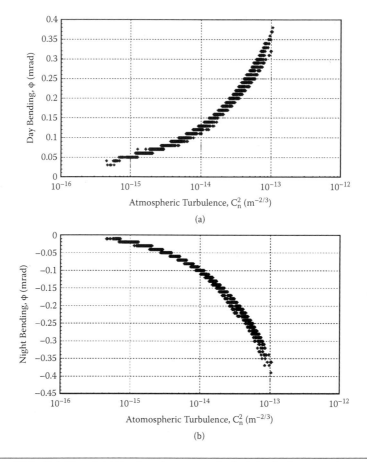

Figure 2.18 Plot of line-of-sight bending values versus refractive index structure parameters modeled by the Wyngaard-Izumu-Collins model [126] for 2 m height above the ground in the Negev area (August 2003): (a) daytime and (b) nighttime data. The measurement data correlation with the model is about 90%, with φ being positive for downward bending. (From S. Bendersky, N. Kopeika, and N. Blaunstein, "Prediction and modeling of line-of-sight bending near ground level for long atmospheric paths," Proc. of SPIE International Conference, San Diego, CA, August 3–8, 2004, pp. 512–522. With Permission.)

Unfortunately, the model introduced in Reference 126 was difficult to evaluate experimentally because of its dependence on wind gradients and temperature gradients. References 79 and 80 consider the refractive index gradient in the general case, without specifying elevation range. In addition, other references deal with this gradient at much higher elevations [129–132].

For application in real-life tactical situations, simple meteorological data values are more usable than gradients of these values. Therefore, the development of a simple model, using direct meteorological measurements, is more desirable for beam bending prediction. However, we limit our work to rather low elevations, from 2 to 15 m. During the period of 3 years of investigations, a semi-empirical model with a high percentage of prediction (about 95%) for line-of-sight bending using simple standard meteorological data was evaluated and developed. This macroscale bending model is presented in the following form:

$$\varphi = \pm \frac{L}{2} C_h \cdot A \cdot \left(C_n^2\right)^B,$$

(2.132)

where

$$A = \alpha \cdot V^{-\beta}$$

(2.133)

$$B = \delta - (\gamma/h)\log V$$

and

$$C_h = \begin{cases} 40.5 & \text{, for} \quad 2 \text{ m} \leq h < 2.5 \text{ m} \\ f(h) & \text{, for} \quad 2.5 \text{ m} \leq h < 5 \text{ m} \\ 1.25 & \text{, for} \quad 5.5 \text{ m} \leq h \leq 15 \text{ m} \end{cases}$$

(2.134)

Here, φ is the line-of-sight bending value (in mrad), L is the distance to the target (in meters), C_n^2 is the refractive index structure parameter (in m$^{-2/3}$), C_n is height factor, and V is wind speed (in m/s). The plus or minus symbols in Equation 2.132 represent day or night bending predictions, respectively, where the bending is downward (positive value of φ) during the day and upward (negative value of φ) during the night. Introduced in Equations 2.133 and 2.134, parameters α, β, δ, γ, and $f(h)$ are the functions of height above the ground. These functions, developed based on measurement data with heights from 2 to 15 m, can be written in the following forms [79]:

$$\alpha = 7.7 \frac{P}{T} h^{-5/6} e^{-h}, \quad \beta = 0.1h + 0.01$$

(2.135)

$$\gamma = 0.073 h^{-0.25}, \quad \delta = 0.73 h^{-0.6}$$

(2.136)

$$f(h) = -0.4267 h^5 + 8.8667 h^4 - 73.767 h^3 + 307.93 h^2$$

$$-647.61h + 553.6$$

(2.137)

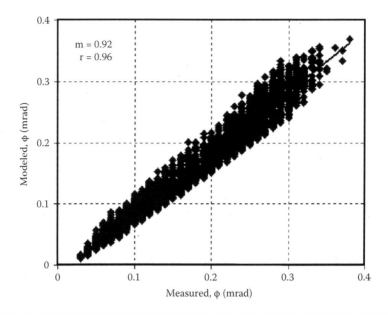

Figure 2.19 Scatterplots of macroscale beam bending model versus measurements for 2 m height in the Negev experimental area during the day; *m* is the slope of regression line and *r* is the Pearson error value. (From S. Bendersky, N. Kopeika, and N. Blaunstein, "Prediction and modeling of line-of-sight bending near ground level for long atmospheric paths," Proc. of SPIE International Conference, San Diego, CA, August 3–8, 2004, pp. 512–522. With Permission.)

where h is the height above the ground (meters), T is the average atmospheric temperature (degrees Kelvin), and P is the average atmospheric pressure (millibars). Equation 2.132 and the parameter α in Equation 2.135 exhibit similar expressions for pressure-temperature ratio (7.7 P/T). Obviously, this relation can be explained by dependence of line-of-sight bending on variations of the refractive index along the optical propagation channels.

Figure 2.19 presents a comparison between measurement and prediction by our empirical model for bending obtained according to Equations 2.132–2.137 for the Negev area for 2 m heights during daytime experiments. The correlation is about 95%. Similar accuracy was found for nighttime and for the Golan experimental data.

Figure 2.20 shows average lines of optical wave bending predicted by the empirical macroscale bending model. The plot is modeled for strong turbulence ($C_n^2 = 1.10^{-13}$ m$^{-2/3}$) and medium turbulence ($C_n^2 = 1.10^{-14}$ m$^{-2/3}$) values versus heights, and wind speeds of 1–5 m/s, with the initial parameters: $L = 4000$ m, $P = 980$ mbar, and $T = 300$ K.

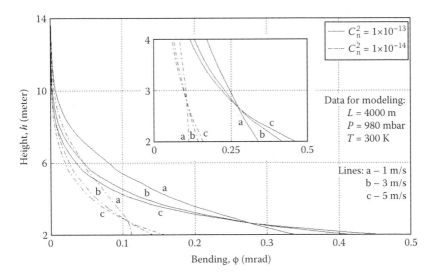

Figure 2.20   Average lines of optical wave bending predicted by the macroscale bending model for different turbulence and wind speed values. (From S. Bendersky, N. Kopeika, and N. Blaunstein, "Prediction and modeling of line-of-sight bending near ground level for long atmospheric paths," Proc. of SPIE International Conference, San Diego, CA, August 3–8, 2004, pp. 512–522. With Permission.)

It can be seen from Figure 2.20 that lines of average bending for different wind speed values between 2 and 3 m high are crossing. Also, for heights above 3 m, the lines have constant direction (decrease with increase of height) and do not change this tendency. This could be explained by the fact that for near-ground altitudes the higher wind speeds increase the heat mass transfer in the atmosphere, causing greater turbulence effects. However, for higher altitudes, increasing wind speeds lead to a decrease in turbulence strength and a decrease of temperature gradient, resulting in beam bending decrease. These comply with common physical phenomena. For heights more than 10 m, the beam bending values became negligible, contrary to phenomena for near-ground levels.

We must note that the proposed model is valid for the following ranges of data: temperature from 9 to 35°C, wind speed from 0.5 to 5.5 m/s, atmospheric pressure from 940 to 990 mbar, and height above the ground from 2 to 15 m. The atmospheric turbulence parameters may be predicted with the model presented in general form by Equation 2.97, according to type of surface and day/night hours [38]. Then, using the macroscale model (Equation 2.97) for prediction of near-ground refractive index structure parameters, the temporal and spatial effects of turbulent atmospheres can be included in

Equation 2.132 for better prediction of optical wave bending for near-surface horizontal propagation.

Finishing this section, we can stress that beam bending near the ground over long atmospheric paths is affected by variations of the refractive index over the optical propagation channel (atmospheric turbulence) and low wind speeds. A variety of the proposed models were examined to explain and predict such line-of-sight bending, but utilization of each is limited only to certain specific conditions.

However, the empirical model over land developed here seems to have much broader applicability. As follows from the results of measurements and corresponding computations according to the model described by Equations 2.132–2.137, such beam bending is negligible for times near sunrise and sunset, when ground and surrounding air temperature are closest. It also exhibits a change of direction from downward bending in daytime to upward bending at nighttime, with respect to real location of remote targets above the ground surface. The diurnal minima near sunrise and sunset are typically absent over water, because water temperature and temperature gradients above water do not vary so much between day and night.

This behavior appears to be related to increased nonuniformity of the atmosphere. Medium wind speeds cause more air mixing and thus give rise to a more uniform atmosphere. This can explain the decrease in beam bending for wind speeds above 4 m/s and the increase in beam bending for lower wind speeds near the ground. Furthermore, at times close to sunrise and sunset, temperature gradients near the ground are minimal, thus causing the air to be more uniform. At midday such gradients are maximum positive and around midnight maximum negative, thus causing the beam bending to be maximum in the downward and upward directions, respectively, at such times when the air is least uniform. It should be noted that increased C_n^2 is also associated with greater nonuniformity of the atmosphere. C_n^2 can be predicted from Equation 2.100 and inserted into Equation 2.132.

References

1. International Telecommunication Union, "Effects of tropospheric refraction on radiowave propagation," ITU-R Recommendation P.834-2, Geneva, 1997.

2. International Telecommunication Union, "Attenuation by hydrometeors, in precipitation, and other atmospheric particles," ITU-R Recommendation P.721-3, vol. V, Geneva, 1990.

3. International Telecommunication Union, "Propagation data and prediction methods required to design of terrestrial line-of-sight systems," ITU-R Recommendation P.530-7, Geneva, 1997.

4. International Telecommunication Union, "Attenuation by atmospheric gases," ITU-R Recommendation P.676-3, Geneva, 1997.

5. International Telecommunication Union, "Specific attenuation model for rain for use in prediction methods," ITU-R Recommendation P.838, Geneva, 1992.

6. International Telecommunication Union, "Attenuation due to clouds and fog," ITU-R Recommendation P.840-2, Geneva, 1997.

7. S. R. Saunders, *Antennas and Propagation for Wireless Communication Systems*, John Wiley & Sons, New York, 1999.

8. D. Deirmendjian, *Electromagnetic Scattering on Spherical Polydispersions*, American Elsevier, New York, 1969.

9. H. M. Nussenzveig and W. J. Wiscombe, "Efficiency factors in Mie scattering," *Phys. Rev. Lett.*, vol. 45, pp. 1490–1494, 1980.

10. W. Zhang, "Scattering of radiowaves by a melting layer of precipitation in backward and forward directions," *IEEE Trans. Antenna Propagat.*, vol. 42, pp. 347–356, 1994.

11. W. Zhang, J. K. Tervonen, and E. T. Salonen, "Backward and forward scattering by the melting layer composed of spheroidal hydrometeors at 5–100 GHz," *IEEE Trans. Antenna Propagat.*, vol. 44, pp. 1208–1219, 1996.

12. H. R. Pruppacher and R. L. Pitter, "A semi-empirical determination of the shape of cloud and rain drops," *J. Atmos. Sci.*, vol. 28, pp. 86–94, 1971.

13. R. K. Crane, "Prediction of attenuation by rain," *IEEE Trans. Commun.*, vol. 28, pp. 1717–1733, 1980.

14. B. R. Bean and E. J. Dutton, *Radio Meteorology*, Dover, New York, 1966.

15. A. Slingo, "A GSM parametrization for the shortwave radiative properties of water clouds," *J. Atmos. Sci.*, vol. 46, pp. 1419–1427, 1989.

16. M. D. Chou, "Parametrizations for cloud overlapping and shortwave single scattering properties for use in general circulation and cloud ensemble models," *J. Climate*, vol. 11, pp. 202–214, 1998.

17. P. S. Ray, "Broadband complex refractive indices of ice and water," *Appl. Opt.*, vol. 11, pp. 1836–1844, 1972.

18. K. N. Liou, *Radiation and Cloud Processes in the Atmosphere*, Oxford University Press, Oxford, England, 1992.

19. N. Blaunstein and C. Christodoulou, *Radio Propagation and Adaptive Antennas for Wireless Communication Links: Terrestrial, Atmospheric, and Ionospheric*, Wiley InterScience, Hoboken, NJ, 2006.

20. R. E. Hufnagle, "Line-of-sight wave propagation through the turbulent atmosphere," *Proc. IEEE*, vol. 56, pp. 1301–1314, 1966.

21. A. Ishimaru, *Wave Propagation and Scattering in Random Media*, Academic Press, New York, 1978.
22. S. M. Rytov, Y. A. Kravtsov, and V. I. Tatarskii, *Principles of Statistical Radiophysics*, Springer, Berlin, 1988.
23. V. I. Tatarskii, *The Effects of the Turbulent Atmosphere on Wave Propagation*, Translations for NOAA by the Israel Program for Scientific Translations, Jerusalem, 1971.
24. L. C. Andrews and R. L. Phillips, *Laser Propagation Through Random Media*, Society of Photo-Optical Instrumentation Engineers, Bellingham, WA, 1998.
25. L. C. Andrews, R. L. Phillips, C. Y. Hopen, and M. A. Al-Habash, "Theory of optical scintillations," *J. Opt. Soc. Am.*, vol. 16, pp. 1417–1429, 1999.
26. L. C. Andrews, R. L. Phillips, and C. Y. Hopen, *Laser Beam Scintillation with Applications*, The International Society for Optical Engineering (SPIE), Bellingham, WA, 2001.
27. P. A. Bello, "A troposcatter channel model," *IEEE Trans. Commun.*, vol. 17, pp. 130–137, 1969.
28. F. G. Stremler, *Introduction to Communication Systems*, Addison-Wesley, Reading, MA, 1982.
29. N. S. Kopeika, *A System Engineering Approach to Imaging*, SPIE Optical Engineering Press, Bellingham, WA, 1998.
30. H. C. van de Hulst, *Light Scattering by Small Particles*, 2nd ed., Dover, New York, 1981.
31. C. F. Bohren and D. R. Huffman, *Absorption and Scattering of Light by Small Particles*, John Wiley & Sons, New York, 1983.
32. F. G. Smith, Ed., *The Infrared and Electro-Optical Systems Handbook*, Vol. 2, *Atmospheric Propagation of Radiation*, SPIE Press, Bellingham, WA, 1993.
33. W. E. K. Middleton, *Vision Through the Atmosphere*, University of Toronto Press, Ontario, 1958.
34. W. J. Wiscombe, "Improved Mie scattering algorithms," *Appl. Opt.*, vol. 19, pp. 1505–1509, 1980.
35. M. E. Thomas and D. D. Duncan, "Atmospheric Transmission," in *The Infrared & Electro-Optical Systems Handbook, Atmospheric Propagation of Radiation*, vol. 2, F. G., Smith, Ed., SPIE Press, Bellingham, WA, 1993.
36. B. Herman, A. J. LaRocca, and R. E. Turner, "Atmospheric scattering," in *The Infrared Handbook*, W. L. Wolfe and G. J. Zissis, Eds., Environmental Research Institute, Ann Arbor, MI, 1989.
37. R. A. McClatchey, R. W. Fenn, J. E. A. Selby, F. E. Volz, and J. S. Garing, *Optical Properties of the Atmosphere*, AFCRL-72-0497, Air Force Cambridge Research Lab, Hanscom AFB, Bedford, MA, 1972.
38. E. P. Shettle and R. W. Fenn, *Models for the Aerosols of the Lower Atmosphere and the Effects of Humidity Variations on Their Optical Properties*, AFGL-TR-79-0214, Air Force Geophysics Lab, Hanscom AFB, Bedford, MA, 1979.
39. R. F. Lutomirski, "Atmospheric degradation of electrooptical system performance," *Appl. Opt.*, vol. 17, no. 24, pp. 3915–3921, 1978.

40. G. C. Mooradian, M. Geller, L. B. Stotts, D. H. Stephens, and R. A. Krautwald, "Blue-green pulsed propagation trough fog," *Appl. Opt.*, vol. 18, no. 4, pp. 429–441, 1979.

41. A. Deepak, A. Zardecki, U. O. Farrukh, and M. A. Box, "Multiple scattering effects of laser beams traversing dense aerosols," in *Atmospheric Aerosols: Their Formation, Optical Properties, and Effects*, A. Deepak, Ed., Spectrum Press, Hampton, VA, 1982.

42. J. W. Strohbehn, "Line-of-sight wave propagation through the turbulent atmosphere," *IEEE Proc.*, vol. 56, pp. 1301–1318, 1968.

43. R. S. Lawrence and J. W. Strohbehn, "A survey of clear-air propagation effects relevant to optical communications," *IEEE Proc.*, vol. 58, pp. 1523–1545, 1970.

44. R. L. Fante, "Electromagnetic beam propagation in turbulent media," *IEEE Proc.*, vol. 63, pp. 1669–1692, 1975.

45. H. T. Yura and W. G. McKinley, "Aperture averaging of scintillation for space-to-ground optical communication applications," *Appl. Opt.*, vol. 22, pp. 1608–1609, 1983.

46. B. E. Stribling, B. M. Welsh, and M. C. Roggemann, "Optical propagation in non-Kolmogorov atmospheric turbulence," *Proc. SPIE*, vol. 2471, pp. 181–196, 1995.

47. A. Zilberman, E. Golbraikh and N. S. Kopeika, "Propagation of electromagnetic waves in Kolmogorov and non-Kolmogorov atmospheric turbulence: Three-layer altitude model," *Appl. Opt.*, vol. 47, pp. 6385–6391, 2008.

48. V. L. Mironov and V. V. Nosov, "On the theory of spatially limited light beam displacements in a randomly inhomogeneous medium," *J. Opt. Soc. Am.*, vol. 67, pp. 1073–1080, 1977.

49. L. A. Chernov, *Wave Propagation in a Random Medium*, Dover, New York, 1967.

50. P. Beckmann, "Signal degeneration in laser beams propagated through a turbulent atmosphere," *Radio. Sci.*, vol. 69D, pp. 629–640, 1965.

51. G. A. Andreev and E. I. Gelfer, "Angular random walks of the centre of gravity of the cross section of a diverging light beam," *Radiophys. Quantum Electron.*, vol. 14, pp. 1145–1147, 1971.

52. J. A. Dowling and P. M. Livingston, "Behavior of focused beams in atmospheric turbulence: Measurements and comments on the theory," *J. Opt. Soc. Am.*, vol. 63, pp. 846–858, 1973.

53. J. R. Dunphy and J. R. Kerr, "Turbulence effects on target illumination by laser sources: Phenomenological analysis and experimental results," *Appl. Opt.*, vol. 16, pp. 1345–1358, 1977.

54. M. A. Kallistratova and V. V. Pokasov, "Defocusing and fluctuations of the displacement of a focused laser beam in the atmosphere," *Radiophys. Quantum Electron.*, vol. 14, pp. 940–945, 1971.

55. T. Chiba, "Spot dancing of the laser beam propagated through the atmosphere," *Appl. Opt.*, vol. 10, pp. 2456–2461, 1971.

56. J. H. Churnside and R. J. Lataitis, "Wander of an optical beam in the turbulent atmosphere," *Appl. Opt.*, vol. 29, pp. 926–930, 1990.

57. V. L. Mironov, *Laser Beam Propagation in the Turbulent Atmosphere*, Nauka, Novosibirsk, USSR, (in Russian) 1981.

58. V. I. Klyatskin and A. I. Kon, "On the displacement of spatially-bounded light beams in a turbulent medium in the Markovian-random-process approximation," *Radiophys. Quantum Electron.*, vol. 15, pp. 1056–1061, 1972.

59. D. H. Tofsted, "Outer-scale effects on beam-wander and angle-of-arrival," *Appl. Opt.*, vol. 31, pp. 5865–5870, 1992.

60. R. J. Cook, "Beam wander in a turbulent medium: An application of Ehrenfest's theorem," *J. Opt. Soc. Am.*, vol. 65, pp. 942–948, 1975.

61. V. I. Klyatskin, *Statistical Description of Dynamical Systems with Fluctuating Parameters*, Nauka, Moscow, 1975.

62. V. E. Zuev, V. A. Banakh, and V. V. Pokasov, *Optics of the Turbulent Atmosphere*, Gidrometeoizdat, Leningrad, 1988.

63. W. L. Wolfe and G. J. Zissis, Eds., "Propagation through atmospheric turbulence," in *The Infrared Handbook*, SPIE Press, Bellingham, WA, 1989.

64. V. Lukin and B. V. Fortes, *Adaptive Beaming and Imaging in the Turbulent Atmosphere*, SPIE Press, Bellingham, WA, 2002.

65. A. I. Kon, V. L. Mironov, and V. V. Nosov, "Dispersion of light beam displacements in the atmosphere with strong intensity fluctuations," *Radiophys. Quantum Electron.*, vol. 19, pp. 722–725, 1976.

66. H. T. Yura, "Short-term average optical-beam spread in a turbulent medium," *J. Opt. Soc. Am.*, vol. 63, pp. 567–572, 1973.

67. A. Zilberman and N. S. Kopeika, "Laser beam wander in the atmosphere: Implications for optical turbulence vertical profile sensing with imaging LIDAR," *J. Appl. Remote Sensing*, vol. 2, 023540 (7 October 2008); DOI:10.1117/1.3008058.

68. C. Rao, W. Jiang, and N. Ling, "Spatial and temporal characterization of phase fluctuations in non-Kolmogorov atmospheric turbulence," *J. Modern Opt.*, vol. 47, pp. 1111–1126, 2000.

69. V. I. Tatarskii, *Wave Propagation in a Turbulent Medium*, McGraw-Hill, New York, 1961.

70. A. S. Monin and A. M. Obukhov, "Basic law of turbulent mixing near the ground," *Trans. Akad. Nauk.*, vol. 24, no. 151, pp. 1963–1987, 1954.

71. V. Thiermann and A. Kohnle, "A simple model for the structure constant of temperature fluctuations in the lower atmosphere," *J. Phys.*, vol. 21, S37–S40, 1988.

72. D. Sadot and N. S. Kopeika, "Forecasting optical turbulence strength on basis of macroscale meteorology and aerosols: Models and validation," *Opt. Eng.*, vol. 31, pp. 200–212, 1992.

73. D. L. Hutt, "Modeling and measurements of atmospheric optical turbulence over land," *Opt. Eng.*, vol. 38, no. 8, pp. 1288–1295, 1999.

74. J. S. Accetta and D. L. Shumaker, Executive Eds., *Environment Research*, SPIE Optical Engineering Press, Bellingham, WA, pp. 157–232, 1993.

75. A. N. Kolmogorov, "The local structure of turbulence incompressible viscous fluid for very large Reynolds numbers," *Reports Acad. Sci.* USSR, vol. 30, pp. 301–305, 1941.

76. C. Y. Young, Y. V. Gilchrest, and B. R. Macon, "Turbulence induced beam spreading of higher order mode optical waves," *Opt. Eng.*, vol. 41, no. 5, pp. 1097–1103, 2002.

77. N. S. Kopeika, I. Kogan, R. Israeli, and I. Dinstein, "Prediction of image propagation quality through the atmosphere: The dependence of atmospheric modulation transfer function on weather," *Opt. Eng.*, vol. 29, no. 12, pp. 1427–1438, 1990.

78. S. N. Bendersky, N. Kopeika, and N. Blaunstein, "Atmospheric optical turbulence over land in middle east coastal environments: Prediction modeling and measurements," *J. Appl. Opt.*, vol. 43, pp. 4070–4079, 2004.

79. S. Bendersky, N. Kopeika, and N. Blaunstein, "Prediction and modeling of line-of-sight bending near ground level for long atmospheric paths," *Proc. of SPIE International Conference*, San Diego, CA, August 3–8, 2004, pp. 512–522.

80. A. Zilberman and N. S. Kopeika, "Aerosol and turbulence characterization at different heights in semi-arid regions," in *Atmospheric Optical Modeling, Measurement, and Simulation*, S. M. Doss-Hammel and A. Kohnle, Eds., *Proc. SPIE*, vol. 5891, pp. 129–140, 2005.

81. J. L. Bufton, "Comparison of vertical profile turbulence structure with stellar observations," *Appl. Opt.*, vol. 12, p. 1785, 1973.

82. R. E. Good, B. J. Watkins, A. F. Quesada, J. H. Brown, and G. B. Loriot, "Radar and optical measurements of C_n^2," *Appl. Opt.*, vol. 21, p. 3373, 1982.

83. J. Vernin, M. Crochet, M. Azouit, and O. Ghebrebrhan, "SCIDAR radar simultaneous measurements of atmospheric turbulence," *Radio Sci.*, vol. 25, p. 953, 1990.

84. R. R. Beland, "Propagation through atmospheric optical turbulence," in *The Infrared and Electro-Optical Systems Handbook*, Vol. 2, F. G. Smith, Ed., SPIE Press, Bellingham, WA, p. 157, 1993.

85. S. S. Oliver and D. T. Gavel, "Tip-tilt compensation for astronomical imaging," *J. Opt. Soc. Am.*, vol. 11, p. 368, 1994.

86. D. Sadot, D. Shemtov, and N. S. Kopeika, "Theoretical and experimental investigation of image quality through an inhomogeneous turbulent medium," *Waves Random Media*, vol. 4, no. 2, pp. 177–189, 1994.

87. C. A. Friehe, J. C. La Rue, F. H. Champagne, C. H. Gibson, and G. F. Dreyer, "Effects of temperature and humidity fluctuations on the optical refractive index in the marine boundary layer," *J. Opt. Soc. Am.*, vol. 65, no. 12, pp. 1502–1511, 1975.

88. V. Thiermann and H. Grassl, "The measurement of turbulent surface layer fluxes by use of bichromatic scintillation," *Boundary-Layer Meteorol.*, vol. 58, pp. 367–389, 1992.

89. J. A. Businger, J. C. Wyngaard, Y. Izumi, and E. F. Bradley, "Fluxprofile relationships in the atmospheric surface layer," *J. Atmos. Sci.*, vol. 28, pp. 181–189, 1973.

90. A. A. M. Holstag and A. P. Van Ulden, "A simple scheme for daytime estimates of the surface fluxes from weather data," *J. Clim. Appl. Meteorol.*, vol. 22, pp. 517–529, 1983.

91. R. B. Stull, *An Introduction to Boundary Layer Meteorology*, Kluwer Academic Publishers, Boston, 1988.

92. N. Ben-Yosef, E. Tirosh, A. Weitz, and E. Pinsky, "Refractive-index structure constant dependence on height," *J. Opt. Soc. Am.*, vol. 69, pp. 1616–1618, 1979.

93. N. S. Kopeika, I. Kogan, R. Israeli, and I. Dinstein, "Prediction of image quality through atmosphere as a function of weather forecast," in *Propagation Engineering*, N. S. Kopeika and W. B. Miller, Eds., *Proc. SPIE*, vol. 1115, pp. 266–277, 1989.

94. D. Dion, and P. B. W. Schwering, "On the analysis of atmospheric effects on electro-optical sensors in the marine surface layer," in *Proc. of the Infrared Information Symposia – 2nd NATO-IRIS Joint Symposium, 25–28 June 1996, London, England*, vol. 41, no. 3, pp. 305–322, ERIM, Infrared Information Analysis Center, Ann Arbor, MI, 1997.

95. R. R. Beland, "Propagation through atmospheric optical turbulence," in *Atmospheric propagation of radiation*, F. G. Smith, Ed., in *The Infrared and Electro-Optical Systems Handbook*, vol. 2, J. S. Accetta and D. L. Shumaker, Executive Eds., Environment Research Institute of Michigan, Ann Arbor, and SPIE Optical Engineering Press, Bellingham, WA, pp. 157–232, 1993.

96. W. B. Miller and J. C. Ricklin, "IMTURB: A module for imaging through optical turbulence," Report ASL-TR-0221-27, U.S. Army Atmospheric Sciences Lab, White Sands Missile Range, NM, 1990.

97. W. J. Stewart, "A propagation model for Gaussian beam waves in clear air turbulence," Report ASL-CR-88-0001-2, U.S. Army Atmospheric Laboratory, White Sands Missile Range, NM, 1988.

98. D. R. Jensen, S. G. Gatham, G. de Leeuw, M. S. Smith, P. A. Frederickson, and K. L. Davidson, "Electro-optical propagation assessment in coastal environments (EOPACE): Summary and accomplishments," *Opt. Eng.*, vol. 40, no. 8, pp. 1486–1498, 2001.

99. J. A. Businger, J. C. Wyngaard, Y. Izumi, and E. F. Bradley, "Fluxprofile relationships in the atmospheric surface layer," *J. Atmos. Sci.*, vol. 28, pp. 181–189, 1973.

100. K. Buskila, S. Towito, E. Shmuel, R. Levi, N. Kopeika, K. Krapels, R. Driggers, R. Vollmerhausen, and C. Halford, "Atmospheric modulation transfer function in the infrared," *Appl. Opt.*, vol. 43, no. 2, pp. 471–482, 2004.

101. S. G. Gathman and M. H. Smith, "On the nature of surf-generated aerosols and their effect on EO systems," in *Propagation and Imaging through the Atmosphere*, L. R. Bissonnette and C. Dainty, Eds., *Proc. SPIE*, vol. 3125, pp. 2–13, 1997.

102. J. Piazzola and S. Despiau, "The vertical variation of extinction and atmospheric transmission due to aerosol particles close above the sea surface in Mediterranean coastal zone," *Opt. Eng.*, vol. 37, pp. 1684–1695, 1998.

103. A. Berk, L. S. Bernstein, and D. C. Robertson, "MODTRAN: A moderate resolution model for LOWTRAN 7," Technical report GL-TR-89-0122, Air Force Geophysics Laboratory, Hanscom AFB, Bedford, MA, 1989.

104. G. R. Ochs, "Measurements of the refractive index structure parameter by incoherent aperture scintillation techniques," in *Propagation Engineering*, N. S. Kopeika and W. Miller, Eds., *Proc. SPIE*, vol. 1115, pp. 107–115, 1989.

105. R. S. Laurence, "A review of the optical effects of the clear turbulent atmosphere," in *Imaging through the Atmosphere*, J. C. Wyant, Ed., *Proc. SPIE*, vol. 75, pp. 2–8, 1976.

106. M. Born and E. Wolf, *Principles of Optics*, Pergamon Press, New York, 1975.

107. A. E. Cole, A. Court, and A. J. Kantor, "Model atmospheres," in *Handbook of Geophysics and Space Environments*, S. Valley, Ed., McGraw-Hill, New York, 1965.

108. J. T. Houghton, *The Physics of Atmospheres*, Cambridge University Press, Cambridge, 1977.

109. K. Krapels, R. Driggers, R. Vollmerhausen, N. S. Kopeika, and C. Halford, "Linear shift invariance of the atmospheric turbulence modulation transfer function for infrared target acquisition models," *Opt. Eng.*, vol. 40, pp. 1906–1913, 2001.

110. J. Laurent, G. Rousset, G. Ferthin, J. F. Carlin, A. Kohnle, V. Thiermman, and M. Drumez, "Comparison between different techniques of turbulence measurements for horizontal path," in *Propagation Engineering*, N. S. Kopeika and W. Miller, Eds., *Proc. SPIE*, vol. 1115, pp. 116–123, 1989.

111. G. R. Ochs, R. R. Bergman, and J. R. Snyder, "Laser beam scintillation over paths from 5.5 to 145 kilometers," *J. Opt. Soc. Am.*, vol. 59, pp. 231–234, 1969.

112. C. A. Friehe, J. C. La Rue, F. H. Champagne, C. H. Gibson, and G. F. Dreyer, "Effects of temperature and humidity fluctuations on the optical refractive index in the marine boundary layer," *J. Opt. Soc. Am.*, vol. 65, no. 12, pp. 1502–1511, 1975.

113. A. A. Holstag and A. P. Van Ulden, "A simple scheme for daytime estimates of the surface fluxes from weather data," *J. Clim. Appl. Meteorol.*, vol. 22, pp. 517–529, 1983.

114. E. J. McCartney, *Optics of the Atmosphere*, Wiley, New York, 1976.

115. M. E. Thomas and D. D. Duncan, "Atmospheric transmission," in *Atmos. Propagat. Radiat.*, F. G. Smith, Ed., in *The Infrared and Electro-Optical Systems Handbook*, vol. 2, J. S. Accetta and D. L. Shumaker, Eds., Environmental Research Institute of Michigan, Ann Arbor, and SPIE Optical Engineering Press, Bellingham, WA, 1993.

116. N. S. Kopeika, "Imaging through the atmosphere," in *Encyclopedia of Physical Science and Technology*, 3rd ed., Academic Press, New York, pp. 661–678, 2001.

117. C. W. Fairall, K. L. Davidson, and G. E. Schachter, "Meteorological models for optical properties in the marine atmospheric boundary layer," *Opt. Eng.*, vol. 26, pp. 847–857, 1982.

118. V. Thiermann and A. Kohnle, "Modeling of optically and IR effective atmospheric turbulence," AGARD-CP-454, Technical paper 19, NATO, Brussels, 1989.

119. J. C. Wyngaard, "On surface layer turbulence," in *Workshop on Micrometeorology*, D. A. Haugen, Ed., American Meteorological Society, Boston, pp. 101–148, 1973.

120. L. D. Landau and E. M. Lifshitz, *Fluid Mechanics*, Pergamon Press, Oxford, 1959.

121. H. A. Panofsky and J. A. Dutton, *Atmospheric Turbulence: Models and Methods for Engineering Applications*, John Wiley & Sons, New York, 1984.

122. H. A. Panofsky, "Spectra of atmospheric variables in the boundary layer," *Radio Sci.*, vol. 4, no. 12, pp. 1101–1110, 1969.

123. H. A. Panofsky, "Spectrum of temperature," *Radio Sci.*, vol. 4, no. 12, pp. 1143–1146, 1969.

124. N. K. Vinnichenco and J. A. Dutton, "Empirical studies of atmospheric structure and spectra in the free atmosphere," *Radio Sci.*, vol. 4, no. 12, pp. 1115–1126, 1969.

125. A. S. Monin and A. M. Yaglom, *Statistical Fluid Mechanics*, MIT Press, Cambridge, MA, 1971.

126. J. C. Wyngaard, Y. Izumu, and S. A. Collins, "Behavior of the refractive index structure parameter near the ground," *J. Opt. Soc. Am.*, vol. 61, no. 12, pp. 1646–1650, 1971.

127. D. H. Tofsted, "Evaluation of the REFRAC refractive ray tracing algorithm," ASL-TR-0242, U.S. Army Atmospheric Sciences Laboratory, White Sands Missile Range, NM, NM88002-5501, 1989.

128. D. H. Tofsted and J. B. Gillespie, "Tactical correction of near surface atmospheric refraction," in *Proc. of Battlefield Atmospherics Conference*, E. Creegan and R. Lee, Eds., 3–6 December 1991, pp. 268–277, U.S. Army Atmospheric Sciences Laboratory, White Sands Missile Range, NM, 1992.

129. S. Y. van der Werf, "Ray tracing and refraction in the modified US1976 atmosphere," *Appl. Opt.*, vol. 42, no. 3, pp. 354–366, 2003.

130. P. E. Ciddor, "Refractive index of air: New equations for the visible and near infrared," *Appl. Opt.*, vol. 35, no. 9, pp. 1566–1573, 1996.

131. J. C. Owens, "Optical refractive index of air: Dependence on pressure, temperature and composition," *Appl. Opt.*, vol. 6, no. 1, pp. 51–59, 1967.

132. C. S. Gardner, "Correction of laser tracking data for the effects of horizontal refractivity gradients," *Appl. Opt.*, vol. 16, no. 9, pp. 2427–2432, 1977.

3

APPLIED ASPECTS OF LIDAR

Atmospheric refractive turbulence and aerosol particles are significant sources of laser-based and imaging system performance degradation. Quantitative estimation and characterization of turbulence and aerosol properties in the atmosphere therefore are very important for such system development and applications (optical communications, laser weaponry, imaging systems, adaptive optics, etc.).

Laser radar, more popularly known as LIDAR (acronym for *light detection and ranging*), is the most powerful technique for active remote sensing of the atmosphere. Ground-based, airborne, and satellite-borne lidar systems are in use [1–10]. Lidar has the desirable feature of being able to measure constitution and optical properties at various heights in the atmosphere.

With more than four decades of use of lidar systems in atmospheric studies, there are a number of reviews and surveys on lidar techniques, system designs and performance, and datasets available in the literature [1–10].

The functional elements and manner of operation of a typical lidar system are schematically illustrated in Figure 3.1. In the simplest form, lidar employs a laser as a source of pulsed energy of useful magnitude and suitably short duration. An intense pulse of optical energy is directed through some appropriate output optics to the atmosphere. As the transmitted laser energy passes through the atmosphere, the gas molecules and particles cause scattering. A small fraction of this energy is backscattered in the direction of the lidar system and is available for detection.

At the receiver, lidar uses a receiving telescope as an optical antenna followed by an optoelectronic detection system. The backscattered energy is collected by means of reflective or refractive optics and transferred to a detector (photomultiplier, avalanche photodiode, etc.). This produces an electrical signal, the intensity of which is proportional to the optical power received. A transimpedance amplifier converts the

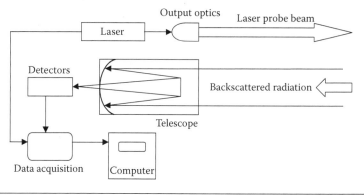

Figure 3.1 Schematic overview of a lidar system.

detector current to a voltage signal suitable for sampling and digitization by an analog-to-digital (A/D) converter and/or a photon counter. The development of very fast waveform digitizers (>20 Msp/s; >14 bit) makes high-resolution data possible.

Because light travels at known velocity, the distance to the scattering volume producing the signal can be uniquely determined from the time interval from pulse transmission to reception. The magnitude of the received signal is determined by the backscattering properties of the atmosphere at successive ranges and by the two-way atmospheric attenuation. Atmospheric backscattering in turn depends on the wavelength of the laser energy used and the number, size, and refractive properties of the particles or molecules illuminated by the incident energy. Thus, the electrical signal from the detector contains information on the presence, range, and concentration of atmospheric scatterers and absorbers. Various forms of presenting and analyzing such signals are available.

Detectability of lidar signals depends mainly on the signal-to-noise ratio (SNR), which is a range-dependent function. The noise includes thermal noise of the electrical circuit involved, shot noise of the photodetector, and noise caused by background optical energy, such as that of solar origin particularly in daytime or thermal emission at all times, depending on wavelength. Because of the high degree of monochromaticity of laser energy, extraneous light can be excluded to a substantial degree by the use of a narrow band filter centered on the laser frequency. Also, because laser energy can be highly collimated, it is possible to direct all the transmitted energy in a narrow beam (typically with a divergence of the order of 0.1–0.6 mrad). Accordingly, the field of view (FOV) of the receiver can be restricted to this angle, thus minimizing background light entering the system.

There are many kinds of lidars, and their classification is not unique. Lidar systems can be classified according to the type of interaction or scattering (elastic or inelastic), instrument configuration, and availability of one, two, or more emission wavelengths. Instrument configuration might be single-ended (or monostatic) or double-ended (or bistatic), depending on whether both the emitter and receiver are located at the same place or not, or coaxial and biaxial, depending on whether the receiver and transmitter optical axes are coincident or not.

The return signal can have an initial rise in the case of an offset configuration, such as when the FOV of the receiver optics increasingly overlaps the path of laser excitation with increasing range. The subsequent fall of the signal is due to the inverse-square decrease with range. This dependence leads to a signal amplitude dynamic range that extends over several decades. Various techniques can be used to compress this range so that the signals are compatible with recording electronics.

In the following sections, applied aspects of lidar use are discussed. The principles of an imaging lidar technique for remote measurements of C_n^2 vertical profile, based on image motion analysis of a secondary source created by a laser beam at a given altitude, are described. Practical examples of imaging lidar experimental results are presented. A short summary of approaches to estimate beam wander statistics for turbulence strength profile retrieval with imaging lidar has been presented in Chapter 2.

Principles of lidar data processing for retrieval of the behavior of the aerosol backscatter fluctuation spectrum (passive scalar turbulence spectrum) are presented and discussed. The results of lidar measurements of aerosol size distribution and aerosol volume and number concentration at the atmospheric boundary layer and free troposphere in the Mediterranean region (Beer-Sheva, Israel) and comparison with models are presented.

3.1 Turbulence Profile Measurement with Lidar

To date, a number of methods have been proposed for remote sensing of C_n^2 vertical profile using lidar techniques. The methods are based on analysis of turbulence-induced scintillation of lidar signals [11–14], image distortion and image motion of secondary sources produced by laser beam [15–19], and phenomena of enhanced backscattering [20, 21].

In References 22 and 23 the experimental validation of the lidar concept of measuring C_n^2 by use of differential image motion (DIM) was presented. The DIM lidar technique employs a pulsed laser and a range-gated imaging system. The laser beam is focused at a selected measurement distance, and light from the scattering volume is received by a telescope with two spatially separated sub-apertures. The variance of the differential motion is directly related to the path-integrated value of C_n^2. By measuring C_n^2 at multiple discrete distances, one can derive the profile of C_n^2 from these measurements.

A short review of recently proposed turbulence strength sensing techniques based on lidar principles was given in Reference 24. However, despite these developments, there is currently no accepted active optical remote sensing technique for directly measuring the C_n^2 profile.

3.1.1 Imaging Lidar Principle

The basic principle involved in lidar sounding of the atmosphere is that atmospheric particles (aerosols and molecules) cause scattering of transmitted energy. Backscattered energy is collected by a suitable optical receiver, which, in turn, gives information about the presence, range, and concentration of various atmospheric scatterers with high range resolution (vertical or horizontal).

In the case of lidar sounding of atmospheric turbulence, aerosol particles appear as a large number of point targets randomly distributed in some volume of space. An imaging lidar system uses this aerosol scattering volume as a virtual monochromatic source that can be located at any altitude by time-gating the receiver (a pulsed transmitter is assumed).

The backscattered optical signal is viewed by a lidar imaging system as a spot or centroid, which is shifted in position for successive lidar returns. The centroid displacements (or dancing) in the focal plane of an imaging system are associated with angle-of-arrival fluctuations of an optical wave in the plane of the receiver aperture. The kind of wavefront distortion that causes centroid dancing is wavefront tilt, when all rays across the aperture arrive at the same angle. This happens when the part of the wavefront that reaches the aperture is relatively small compared to the size of turbulence eddies.

To determine the structure parameter C_n^2 from imaging lidar measurements, the general approach is to use the temporal changes of the center of gravity of the image. The standard deviation, σ, from the location of image center of gravity is a measure of the fluctuations of the angle of arrival. Because image motion usually takes place in the focal plane of the imaging system (for relatively large object distances), the mean square displacement, $<\sigma^2>$, of the image can be connected with angle-of-arrival variance, $\langle \varphi_{AA}^2 \rangle$, using the approximation $\langle \sigma^2 \rangle = f_{\text{eff}}^2 \cdot \langle \varphi_{AA}^2 \rangle$, where f_{eff} is the receiver effective focal length.

The wave angle-of-arrival variance in the plane of the receiver aperture is directly related to the path-integrated value of C_n^2 as

$$\left\langle \varphi_{AA}^2 \right\rangle \sim \bar{C}_n^2 = \int_0^L C_n^2(z) \cdot Q(z,L) dz \tag{3.1}$$

where $Q(z, L)$ is the optical weighting function for the path and L is the path length. This function describes the relative effectiveness of each portion of the propagation path for producing wavefront tilt (or angle-of-arrival fluctuation).

By measuring the path-integrated values of C_n^2 at different discrete distances and using an inversion algorithm for the corresponding integral equation, one then can derive the C_n^2 profile from such measurements. The difficulties in implementation of this concept for refractive turbulence sensing arise from the fact that the lidar applications are concerned with a roundtrip path: from the lidar to the scattering volume (secondary source) and back to the receiver.

The uplink propagation produces turbulence-induced beam centroid wander at some selected altitude. The downlink propagation of the backscattered wavefront produces the angle-of-arrival (or wavefront tilt) fluctuations that change the position of the illuminated scattering volume (secondary motionless source) in the image plane of the receiver. Thus, the secondary source displacements in the image plane of the receiver contain two components corresponding to the random motion of the transmitted beam and the motion of a secondary source imaged by the receiver.

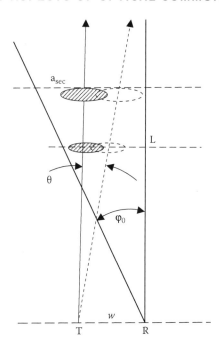

Figure 3.2 Configuration for analysis: T is transmitter, R is receiver, L is height of scattering volume, a_{sec} is beam centroid (secondary source cross-section) diameter, φ_0 is receiver FOV, θ is angular beam deviation from the nominal pointing direction produced by refractive turbulence, and w is baseline distance between transmitter and receiver.

For the sake of convenience, the problem is defined by the configuration shown in Figure 3.2. The system is configured as an off-axis monostatic imaging lidar.

The secondary source is created by the laser transmitter placed near the receiver. The wavefront tilt (or angle-of-arrival) variance at the receiver plane can be written in general form as [25, 26]

$$\left\langle \varphi_{AA}^2 \right\rangle = \frac{\left\langle \rho_{im}^2 \right\rangle}{f_{eff}^2} = \left\langle \varphi_{LB}^2 \right\rangle + \left\langle \varphi_{SS}^2 \right\rangle - 2 \left\langle \varphi_{LB} \varphi_{SS} \right\rangle \qquad (3.2)$$

where $\left\langle \varphi_{LB}^2 \right\rangle$ is the variance of random angular displacements of the beam centroid on the upward path, $\left\langle \varphi_{SS}^2 \right\rangle$ is the wavefront tilt variance of the reflected wave from the secondary motionless source at the receiving aperture, $\left\langle \varphi_{LB} \varphi_{SS} \right\rangle$ is the mutual correlation between fluctuations of a random angular displacement of transmitted beam centroid at plane $z = L$ and backward wavefront tilt at receiver plane $z = 0$, ρ_{im} is the root mean square (RMS) centroid position imaged in the focal plane of the receiver, and f_{eff} is the effective focal length of the receiving system. Note that the atmospheric transmittance and aerosol scattering do not influence wavefront tilt variance. It is assumed also that the transmitter has perfect beam pointing.

This concept is nearly related to the tilt sensing technique for adaptive optics correction, where some artificial reference source provides information concerning the distribution of wavefront fluctuations in the propagation path and is used for wavefront sensing. In adaptive astronomy, an artificial reference source usually is called a laser guide star (LGS) or laser beacon [25].

There are two main schemes of LGS generation: monostatic and bistatic. The monostatic scheme can include different configurations: on-axis or coaxial transmitter/receiver system, and off-axis, where the transmitter and receiver are placed a small distance apart (see Figure 3.2).

Several techniques for wavefront tilt measurement with an LGS have been proposed and analyzed [25–32]. In the monostatic or bistatic configuration it is usually assumed that a focused beam forms the LGS in the atmosphere at some altitude and then the image of the LGS is recorded. In this case the secondary source can be assumed to be point one. However, in the monostatic scheme with a focused beam, the fluctuations on direct and reversed paths are entirely correlated according to path reciprocity that reduces or eliminates the wavefront tilt at the receiver. It has been pointed out [26] that a conventional laser beacon is unable to sense a full aperture tilt and can be used to measure only the higher order wavefront distortions.

In Reference 33 it was shown that the reciprocity of propagation paths might be broken down by using a transmitter and receiver with different diameters, D_T and D_R, respectively, $D_R > D_T \gg (L/k)^{1/2}$, where k is the wavenumber, or by using a divergent beam if its effective size at a laser beacon altitude coincides with the radius of a receiving telescope.

In the bistatic configuration, the ray trajectories in the transmitted beam and the reflected wave do not coincide with each other, and statistically independent paths can be assumed (e.g., $\langle \varphi_{LB} \, \varphi_{SS} \rangle \equiv 0$) that eliminate tilt reciprocity.

The influence of atmospheric refractive-index fluctuations on the propagation of an optical wave when the wave traverses the same region of the atmosphere twice was previously investigated. The properties were determined experimentally and theoretically by correlations between an incident wave and a reflected one traversing the same inhomogeneities. The random motion of the image of the aerosol scattering volume as well as that of different targets illuminated by the laser beam has been analyzed [34–39]. It was found that it is possible to realize an optical scheme where mutual correlation for fluctuations in direct and backward beams takes some intermediate value.

In the present description, the off-axis monostatic lidar configuration as shown in Figure 3.2 with a collimated or slightly divergent beam is considered and analyzed.

It can be noted that an additional problem exists in the proposed measurement scheme. After propagation inside the optical active turbulence layer, some form of saturation of variance of angular displacements with path length increase can be observed. It follows from the fact that phase effects (wavefront tilt) are dominated by strong turbulence that occurs near the receiving or transmitting aperture because of their low elevations, and in the atmospheric boundary layer. If the transmitter is near the Earth's surface, measurements of backscattering above the boundary layer will include these large distortions, and contributions from above the boundary layer can be swamped by the larger contribution from below that can result in errors in C_n^2 vertical profile retrieval.

The mutual correlation between the random displacements of beam centroid and its image has been investigated previously [27, 28, 39, 40] for the case where the beam and the backscattered wave propagate along nearly the same optical paths.

In Reference 41 the correlation between displacements of outgoing beam and angle of arrival of backscattered wave at the receiver, pertaining to Figure 3.2, has been considered. Also, the approximate expression for angle-of-arrival variance at the receiver plane for roundtrip propagation has been obtained where the correlation between angular displacements over the upward and downward paths was taken into account. The mean square of the angle of arrival is given by [41]

$$\left\langle \varphi_{AA}^2 \right\rangle = \left[\frac{4.07}{a_0^{1/3}} + \frac{5.65}{D^{1/3}} - 9.14 \cdot {}_1F_1\left(\frac{1}{6}, 1; -\frac{w^2}{D^2 + a_0^2} \right) \cdot \left[D^2 + a_0^2 \right]^{-1/6} \right]$$

$$\cdot \int_0^L dz C_n^2(z) \left(1 - \frac{z}{L} \right)^2 \tag{3.3}$$

where a_0 and D are the initial beam radius (e^{-1} intensity radius of the Gaussian beam at the source) and diameter of the receiver, respectively, and w is the baseline distance between the transmitter and receiver (see Figure 3.2). The outer turbulence scale is assumed to be much larger than the outgoing beam size, and the turbulence-induced and diffractive beam widening is negligible.

Assuming that there is no correlation between φ_{LB} and φ_{SS}, the mean square of the angle of arrival is given by [41]

$$\left\langle \varphi_{AA}^2 \right\rangle = \left[\frac{4.07}{a_0^{1/3}} + \frac{5.65}{D^{1/3}} \right] \cdot \int_0^L dz C_n^2(z) \left(1 - \frac{z}{L} \right)^2 \tag{3.4}$$

Note that in all cases it is assumed that the secondary source size, a, can be approximated as a point source ($a \le (\lambda L)^{1/2}$). The lidar system measures $\langle \varphi_{AA}^2 \rangle$ from the gated return at some altitude L (see Figure 3.2). Following Equations 3.3 and 3.4, it can be written as an integral along the optical path:

$$\left\langle \varphi_{AA}^2(L) \right\rangle = const \cdot \int_0^L C_n^2(z) \cdot Q(L, z) dz \tag{3.5}$$

where $Q(L, z)$ is the path-weighting function.

3.1.2 Practical Considerations

Figure 3.3 illustrates the possible optical design of an imaging lidar system. The major components include the telescope, laser with beam-expanding optics, beam jitter sensor, imaging module, and control module (delay generator and PC).

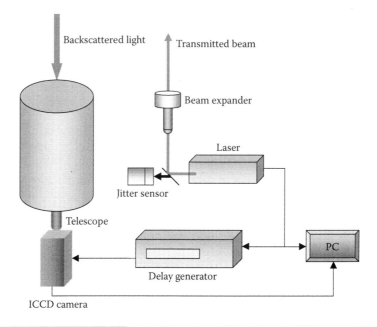

Figure 3.3 Imaging lidar system setup, off-axis monostatic configuration. ICCD = intensified charge-coupled device. (From A. Zilberman and N. S. Kopeika, "Laser beam wander in the atmosphere: Implications for optical turbulence vertical profile sensing with imaging LIDAR," *J. Appl. Remote Sensing*, vol. 2, 023540, 1 January 2008. With Permission.)

The laser beam is expanded and transmitted vertically into the atmosphere without focusing at the altitude of the selected atmospheric scattering region. Before entering the beam expander, the beam direction is sampled by the jitter sensor. In the angle-of-arrival statistics calculation, it is assumed that the transmitter is ideal and has perfect beam pointing. In real-world laser systems, the beam-pointing instability can reach 30–40 μrad RMS, which can be up to an order of magnitude more than turbulence-induced wander. The jitter sensor (quadrant or charge-coupled device [CCD]-based detector) measures the propagation angle of the laser beam as it exits the laser. This information is stored and then can be used to null out effects of laser-induced beam jitter.

The backscattered light from the scattering volume (secondary source), after being focused by the telescope, is collimated, passed through the interference filter, and detected by an imaging module and stored. The expected angular image motion can be calculated on the basis of Equations 3.3 and 3.4.

An intensified CCD (ICCD) camera can be used as an image detector placed at the focal plane of the telescope. The ICCD has a high

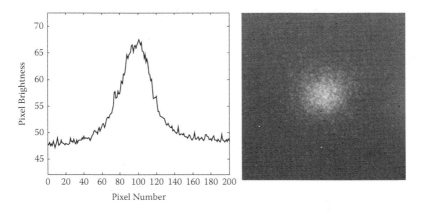

Figure 3.4   Laser beam spot from 9 km altitude imaged by a receiving system; 15 µs exposure time [42].

gain and changeable time gating (exposure time). To allow for synchronization between laser firing and signal detection, a fixed pulse is derived from the Q-switch that is made to trigger the time-delay generator and ICCD. Thus, a laser pulse is range gated at the receiver, which allows it to obtain a series of time- and altitude-resolved images of the atmospheric scattering volume.

The scattering volume (or beam centroid) is viewed by the imaging system as a spot, which is shifted in position for sequences of lidar returns. To estimate the path-weighted value of C_n^2, the general approach is to use the temporal changes of the center of gravity of the image. The mean square location $\langle \sigma^2 \rangle$ of the image center of gravity is a measure of the fluctuations of the angle of arrival $\langle \varphi_{AA}^2 \rangle$. For calculation of angle-of-arrival variance as a function of height, a set of images from each altitude should be used. Figure 3.4 shows an example of an image of beam centroid from 9 km height elevation and mean image distribution brightness.

Using a CCD camera at the receiving telescope output, one can derive the variance of the angle of arrival of an incoming wave from $\langle \sigma^2 \rangle$ as a function of pixel size, d (m/pixel), and effective focal length of the system, f_{eff} (m), as follows:

$$\left\langle \varphi_{AA}^2 \right\rangle = \frac{[d^2 \cdot \langle \sigma^2 \rangle]}{f_{\mathrm{eff}}^2} \qquad (3.6)$$

The centroid position for sequence of images is calculated from the estimation of image center of gravity, and then the variance $\langle \sigma^2 \rangle$ is determined. The value of $\langle \varphi_{AA}^2 \rangle$ is calculated from Equation 3.6.

Images are encoded as two-dimensional (2D) arrays of pixel values [i.e., integers between 0 and 255 (gray scale)] describing the darkness of the corresponding pixel for 8-bit CCD dynamic range.

We denote by $\mathbf{x} = \{x_i\}$ and $\mathbf{y} = \{y_i\}$ the vectors of all pixel positions and by $I(x_i, y_i)$ the discrete function defining the pixel values of the image (gray level at a pixel changed from 0 to 255) at coordinate $[x_i, y_i]$ within an $M \times N$ image.

To calculate the mean square of the image center-of-gravity displacements, the precise centroid position needs to be obtained. The technique that is used to estimate the centroid position follows from image moment analysis where the first spatial moment of the image distribution corresponds to the center-of-mass (centroid) position.

For the one direction case (along rows or columns), when the image distribution is sampled in a discrete set of points with spatial coordinates $\{x_i, y_i\}$, the centroid (center-of-mass) position is given by

$$X_c = \frac{\sum_j \sum_i x_i I(x_i)}{\sum_j \sum_i I(x_i)} \tag{3.7}$$

where $I(x_i)$ is the intensity or gray level at point x_i of the image. For 2D calculations, the centroid position, Y_c, for the y direction is calculated from Equation 3.7.

From a set of images, the variance of image center of mass positions can be defined as

$$\sigma^2 = \frac{1}{J-1} \sum_{i=1}^{J} \left\{ X_c^2 + Y_c^2 \right\} \tag{3.8}$$

where J is the number of images and $<X_c> = 0$ and $<Y_c> = 0$.

3.1.3 Lidar Inaccuracy

The lidar SNR required for measuring the profile of angle-of-arrival variance with sufficient accuracy to infer useful information about C_n^2 vertical profile depends on details of the detector and optical system. A generalized criterion for required SNR is that the system be able to detect weak backscattered signals and sense the change in image position of the scattering volume of interest. Low SNR in the image can be a problem for high-altitude sensing, where aerosol particle density is small. The image of scattering volume from these altitudes consists of separate photon counts, which can result in center-of-gravity calculation errors.

The problem of determining the centroid of a digital image has been discussed in References 43–47. It was found that the RMS wavefront tilt measurement error can be estimated to be $\Delta\varphi \approx b/SNR$, where b is the effective spot size in the image and SNR is the signal-to-RMS noise ratio. In general, the centroiding accuracy depends on SNR, window size (CCD area), pixel size, and image spot size (or ratio of window size to spot size), which are important design considerations.

Errors in mean square centroid displacement estimation and, thus, in angle-of-arrival variance include inaccuracy in center-of-mass estimation as well as errors caused by a limited number of images. To determine the accuracy of wavefront tilt measurement from centroid position calculation (centroiding), analysis of received images needs to be performed.

Figure 3.5 shows a 3D plot of the short-exposure image of an LGS received by lidar. An undesirable background level exists in the experimental data. Its sources can be dark currents, background light, unbalanced CCD and frame-grabber zero levels, and so on. Also, the measurements are corrupted by noise (photon noise and readout noise of the CCD). Using the CCD as a photon detector integrated into the imaging device, the light-intensity distribution I_{ij} in discrete samples (pixels) is obtained. The image intensity distribution can be presented as

$$I_{ij} = I_{0ij} + n_{ij} \tag{3.9}$$

where I_{0ij} is the noiseless intensity in pixel $\{i, j\}$; n_{ij} is the noise intensity; i, j is the pixel coordinate; and $i, j = 1 \ldots W$, where W is the window size (image size). An additive noise $\mathbf{n}$ consists of all the disturbances, including background and sensor noise.

Considering that the recorded intensity over each pixel can be modeled with Equation 3.9, the centroid can be expressed as [42]

$$X_c = \frac{\sum_i \sum_j x_j I_{0j}}{\sum_i \sum_j (I_{0j} + n_j)} + \frac{\sum_i \sum_j x_j n_j}{\sum_i \sum_j (I_{0j} + n_j)} \tag{3.10}$$

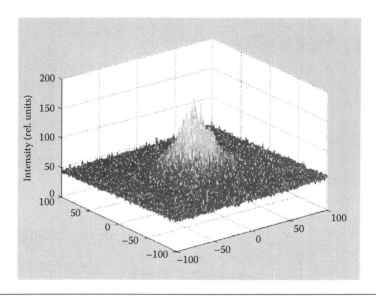

Figure 3.5 Example of a 3D plot of a received LGS image [42].

The noise and background level in the data, $n(n_0, \sigma)$, reduce accuracy in centroid computation.

The noise statistics can be estimated by analyzing the intensity distribution for a flat field (in a homogeneously illuminated background section of the image). The histogram of the intensity distribution has a Gaussian-like statistical dispersion (see Figure 3.6).

Figure 3.7 shows a probability plot of a flat-field intensity distribution. The plot is linear, indicating that the sample can be modeled by a normal distribution.

Thus, the noise can be modeled by a Gaussian distribution function with space variant mean and variance: $\mathbf{n} \sim \mathbf{N}(n_0, \sigma)$. From these experimental data, the maximum likelihood estimates (MLEs) for the parameters of a normal distribution were calculated to be $n_0 \sim 50$ and $\sigma \sim 3.1$, corresponding to mean background level and RMS noise level (the noise standard deviation). The SNR is calculated as [42]

$$SNR = \frac{\sqrt{W} \cdot [\max(< I >_X) - n_0]}{\sigma} \qquad (3.11)$$

where max $(<I>_X)$ is the maximal value of image brightness distribution averaged along the X coordinate, n_0 is the background brightness

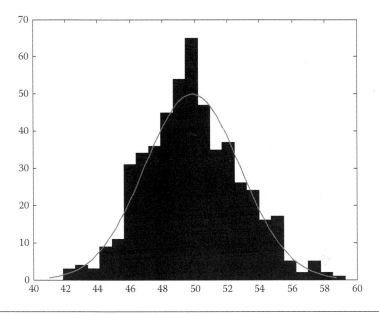

Figure 3.6   Flat-field intensity distribution [42].

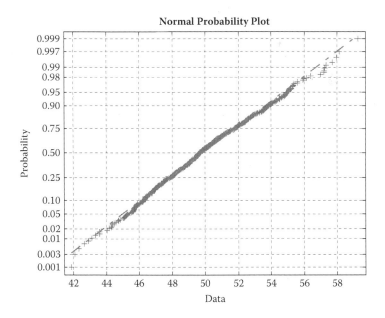

Figure 3.7   Probability plot of a flat-field intensity distribution. The plot is linear, corresponding to a normal distribution [42].

value in the image, σ is the noise standard deviation, and W, as before, is the number of averages (window size).

As mentioned previously, the straightforward centroid position computation, in the form of Equation 3.7, yields errors if there is some background intensity level (as is usually the case in practice). For accurate centroid computation, the threshold value must be chosen in order to reject pixels with small SNRs. An analytical description of the interaction between centroiding and thresholding applied over an intensity distribution corrupted by additive Gaussian noise was presented in Reference 46.

To remove truncation error, the ratio of the image size (W) to spot size at e^{-1} intensity level needs to be larger than 3, or W/I_{FWHM} must be larger than 4, where I_{FWHM} is the full width of half maximum of spot intensity distribution [44]. The truncation error occurs because the spot has a certain width on the detector, and limiting the region of integration involves truncating part of the spot.

Figure 3.8 shows the result of thresholding (median method) of the intensity distribution averaged along rows of the image, $<I>_X$, for lidar-received images. The thresholding does not change the centroid position. The behavior of the errors in the centroid position as a function of the SNR (dB) is shown in Figure 3.9 for different W/I_{FWHM} ratios.

Let us consider $S_0 = W/I_{\text{FWHM}}$. The analysis shows that accuracy in centroid calculations depends on SNR and S_0 as $\delta C \sim 1/\text{SNR}(S_0)$, where SNR changes with S_0 and δC corresponds to error in the centroid position. Note that SNR (dB) was calculated as $\text{SNR}_{dB} = 10 \log(\text{SNR})$. The optimal value of S_0 corresponds to the interval $3 < S_0 < 6$.

A 1-pixel error will correspond to wavefront tilt estimation error $\delta\varphi \approx 2$ μrad for $f_{\text{eff}} = 10$ m and $d = 24$ μm pixel size, which may correspond to about 10% inaccuracy in the angle-of-arrival lidar calculation. Thus, Equation 3.6 can be written a in different form, as follows:

$$< \hat{\varphi}_{AA}^2 > = < \hat{C}^2 > \cdot \frac{d^2}{f_{\text{eff}}^2} \tag{3.12}$$

where $\hat{\varphi}, \hat{C}$ are the estimates of true values.

Figure 3.10 shows an example of a measured angle-of-arrival variance as a function of altitude.

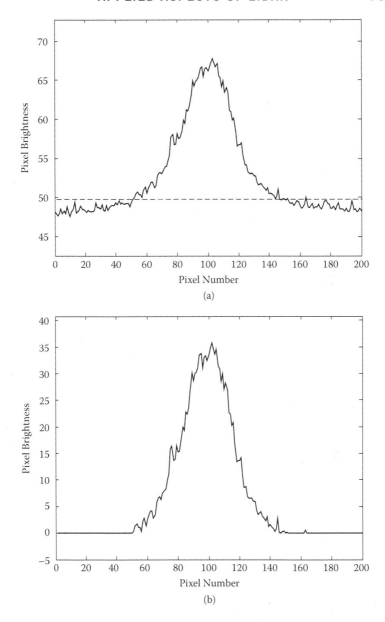

Figure 3.8 Thresholding of the image intensity distribution of lidar-received signal averaged along one axis, $<I>_\chi$ [42]; SNR = 16 dB; (a) before thresholding, (b) after thresholding. Dashed line corresponds to the median level of mean intensity distribution.

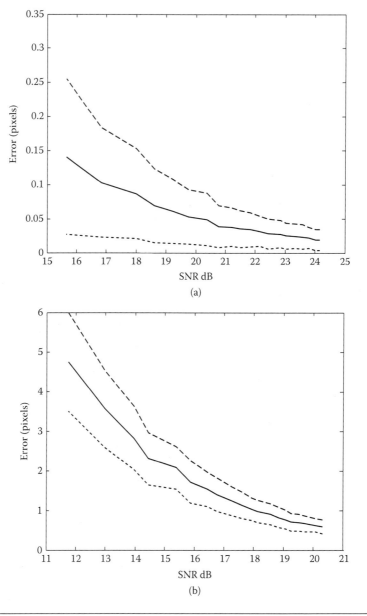

Figure 3.9 The accuracy in centroid calculation vs. SNR for different W/I_{FWHM} ratios [42]: (a) $W/I_{FWHM} = 4$, (b) $W/I_{FWHM} = 9$. Dashed line is the 1σ values.

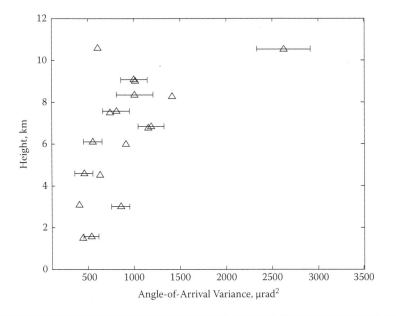

Figure 3.10   Angle-of-arrival variance versus height measured on 3 October 2004, 11:00 p.m. (local time) Beer-Sheva (Israel) [42]; 75 m difference at neighboring altitude elevations; 2 Hz laser repetition rate. Error bars indicate an accuracy of $<\varphi_{AA}^2>$ estimation.

3.1.4 C_n^2 Retrieval Technique

If C_n^2 is assumed to be constant over the interval $[L_{j-1}, L_j]$, Equation 3.5 can be rewritten as a sum of integrals over the range gates:

$$\left\langle \varphi_{AA}^2(L_i) \right\rangle = const \cdot \sum_{j=1}^{i} C_n^2(L_j) \cdot \int_{L_{j-1}}^{L_j} Q(L_i, z)dz \qquad (3.13)$$

This permits writing the result in matrix form as

$$\frac{\left\langle \varphi_{AA}^2(L_i) \right\rangle}{const} = \sum_{j=1}^{i} C_n^2(L_j) \cdot Q_{i,j} \qquad (3.14)$$

or, in terms of vectors **g** and **f** and matrix **Q**,

$$\mathbf{g} = \mathbf{Q} \cdot \mathbf{f} \qquad (3.15)$$

where **g** is the measured data vector, and **f** is the unknown function. In general, the matrix **Q** has elements that satisfy $Q_{ij} = 0$ for $j > i$.

Thus, all the upper-half matrix elements are zero, and the formulation gives a triangular system of equations. The solution of this system can be written as

$$C_n^2(L_j) = \frac{\left[\varphi_{AA}^2(L_i) - \sum_{j=1}^{i-1} C_n^2(L_j) \cdot Q_{ij}\right]}{Q_{ii}}, \quad i = 2, \ldots, N. \qquad (3.16)$$

where $C_n^2(L_1) = \varphi_{AA}^2(L_1)/Q_{11}$ and N is the number of altitude elevations. The same result can be obtained by direct inversion of Equation 3.15:

$$\mathbf{f} = \mathbf{Q}^{-1} \cdot \mathbf{g} \qquad (3.17)$$

However, in practice, these equations generally may not necessarily lead to useful results, because vector $\mathbf{f}$ oscillates and even becomes negative. The solution process is known to be ill-posed, meaning that there is more than one solution, and small noise in the observed data can significantly deteriorate the solution.

Because experimental measurement errors are involved, Equation 3.15 can be revised in a more correct form as

$$\mathbf{g} = \mathbf{Q} \cdot \mathbf{f} + \boldsymbol{\varepsilon} \qquad (3.18)$$

where $\boldsymbol{\varepsilon}$ is measurement errors in terms of an error vector.

To obtain the required turbulence strength profile from the measured data, Equation 3.18 must be solved for the function $\mathbf{f}$ by suitable inversion method. The inversion of an ill-posed problem has been studied by different methods [48–52].

Typically, some kind of regularization (regularization method) is employed to obtain reasonable approximations to the true data. Following this method, the solution of Equation 3.18 is of the form [53]

$$\mathbf{f} = (\mathbf{Q}^T\mathbf{Q} + \chi\mathbf{H})^{-1}\mathbf{Q}^T\mathbf{g} \qquad (3.19)$$

where $\mathbf{H}$ is the smoothing matrix [54], χ is the regularization parameter (non-negative Lagrange multiplier) selected so that all the elements of $\mathbf{f}$ are positive [55, 56], and $\mathbf{Q}^T$ is the transpose of $\mathbf{Q}$.

The matrix $\chi\mathbf{H}$ serves to take out the oscillation in $\mathbf{f}$ and, for $\chi = 0$, Equation 3.19 reduces to Equation 3.17. The solution in this technique depends on choosing a regularization parameter, χ. If χ is too small, the solution vector $\mathbf{f}$ will oscillate and can have negative values. If χ is large, the solution becomes smoothed over.

Chahine [57, 58] proposed a nonlinear iterative method to determine temperature profiles of the atmosphere from measurements of its emerging radiance as a function of frequency. The main advantages of this method are that no tuning of external parameters is needed; i.e., no constraints are imposed on the solutions that are always positive.

The inversion method is an iterative scheme that leads to obtaining the next iteration distribution once the previous distribution is known. The basic idea is to find a function whose values have minimum deviation from the measured ones.

Let us suppose that the C_n^2 profile has been recovered after p iterations: ($K_i^P = C_n^2(z_i)$, $i = 1, ..., N$). This profile gives rise to a sequence of signals, $\alpha_{p,\text{calc}}(z_i) = \alpha^2(z_i)[\langle\varphi_{AA}^2\rangle \equiv \alpha^2(z_i)]$, calculated according to Equations 3.12 and 3.14, which will be different from the sequence of the measured signals, $\alpha_{\text{meas}}(z_i)$. To find a better fit for the profile at the $p + 1$ iteration, the scheme corrects it in the following way:

$$K_i^{p+1} = K_i^p \cdot \frac{\alpha_{\text{meas}}(z_i)}{\alpha_{p,\text{calc}}(z_i)} \tag{3.20}$$

The convergence and reliability of the method is estimated by calculating the root mean error (RME), which describes the relative RMS deviation of the retrieved signals, $\alpha_{\text{calc}}(z_i)$, from the measured signals, $\alpha_{\text{meas}}(z_i)$:

$$\text{RME} = \left\{ \frac{1}{N} \sum_{i=1}^{N} \frac{[\alpha_{\text{meas}}(z_i) - \alpha_{\text{calc}}(z_i)]^2}{[\alpha_{\text{calc}}(z_i)]^2} \right\}^{1/2} \tag{3.21}$$

where N is the number of altitude elevations.

A modification to this method was introduced by taking an additional weighting coefficient, given by

$$\bar{Q}_i = \sum_j \sum_i Q_{i,j}, \quad j = 1,...,i \tag{3.22}$$

and Equation 3.14 is modified as follows:

$$\alpha^2(z_i) = \alpha_{\text{calc}}(z_i) = \frac{\sum_{j=1}^{i} C_n^2(z_j) \cdot Q_{i,j} \cdot \bar{Q}_i}{\bar{Q}_i}. \quad i = 1,...,N. \tag{3.23}$$

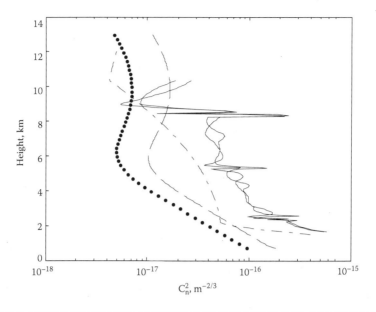

Figure 3.11 Retrieved C_n^2 profiles for two nights — 27 August 2003 (solid line) and 28 August 2003 (dotted line), 11:30 p.m.–1:00 a.m. local time — and profiles calculated from models: the AFGL CLEAR I night model (dash-dot) for the New Mexico desert in summer; modified version of the HV model that is typical for Mauna Kea, Hawaii [31] (dots); the HV-21 model that is valid in Albuquerque, New Mexico (dash). (From A. Zilberman and N. S. Kopeika, "Aerosol and turbulence characterization at different heights in semi-arid region," in *Atmospheric Optical Modeling, Measurement, and Simulation, Proc. SPIE*, no. 5891, pp. 129–140, 2005. With Permission.)

This weighted value of $\alpha^2(z_i)$ is used in the iteration scheme (Equation 3.20). We can call this a modified Chahine approach.

By use of Equation 3.23, a good match between the calculated and measured signals is achieved very quickly. The iteration stops when the RME reaches the minimum value. The inversion procedure is very stable with respect to the number of iterations. The zero-iteration profile of C_n^2 can be chosen as a constant. The final result is independent of this initial guess.

Figure 3.11 shows restored C_n^2 profiles for two nights — 27 August 2003 (solid line) and 28 August 2003 (dotted line) — and comparison with C_n^2 profiles calculated from models.

3.2 Lidar Research of Passive Scalar Field Fluctuations

As mentioned in Chapter 1, the Obukhov-Kolmogorov (O-K) type of refractive index fluctuation spectrum in the atmosphere (Equation 1.35) corresponds also to fluctuations of the passive scalar field (temperature, aerosols, etc.). Aerosol particles suspended in the air can be considered as a passive scalar for particle radius <5 μm and low concentration. The aerosol inhomogeneities represent variations in space and time of aerosol microphysical characteristics, such as concentration and size distribution. In general, these variations are due to the atmospheric turbulent mixing and the stratification of meteorological parameters.

Experimental study of the behavior of the atmospheric passive scalar turbulence spectra can be based on measurements of the intensity fluctuations of lidar signals scattered by aerosol inhomogeneities. Fluctuations of aerosol particulate concentration are responsible for variations in the scattered intensity of light propagating in turbulent media. In lidar applications, the intensity (or power) fluctuations correspond to backscattering coefficient fluctuations. Thus, the lidar return signal fluctuations can be related to aerosol inhomogeneity variations resulting from the influence of atmospheric processes, and the behavior of atmospheric turbulence spectra in various turbulent scales can be estimated. The possibility of such studies was proved in References 60–62 for different atmospheric conditions.

3.2.1 Lidar Method for Turbulence Spectrum Estimation

For a sequence of laser shots, temporal fluctuations of the power of a lidar-received signal from the distance z elastically scattered by aerosols in a non-turbid atmosphere can be described by the lidar equation under single-scattering approximation as follows [1]:

$$P(z,t) = B \cdot \beta_\pi(z,t) \cdot z^{-2} \cdot T^2 \qquad (3.24)$$

where $P(z, t)$ is the received power at moment t from distance z; $\beta_\pi(z)$ is a volume backscatter coefficient; T is atmospheric transmission; and B is a system constant. Assuming that the particle number density fluctuations in the scattering volume cause spatial-temporal variations of $\beta_\pi(z)$, the volume backscatter coefficient can be written as [62]

$$\beta_\pi(z,t) = N(z,t) \int \sigma_\pi(r) f(r) dr \qquad (3.25)$$

where $\sigma_\pi(r)$ is a backscattering cross-section of particles with the radius r; $f(r)$ is the distribution function of particle sizes; and $N(z, t)$ is a spatial-temporal concentration field.

If the path transmittance does not change significantly during the experiment, the temporal variations of a returned lidar signal can be represented as $P(z, t) \sim \beta_\pi(z, t) \sim N(z, t)$; i.e., to a first approximation it is proportional to aerosol concentration fluctuations in the scattering volume.

The power spectral density (PSD), $W(z, f)$, of the laser returned signal from the altitude z corresponds to the PSD of the passive scalar, $\Phi_n(z, f)$, where f is the temporal frequency. Performing measurements with different laser repetition rates (sampling times), it is possible to estimate the contribution of various turbulent scale sizes in the spectrum.

Lidar measurements of the returned signal power fluctuations are performed in the time (frequency) domain. However, in interpreting the results, it is important to examine the behavior of $\Phi(z, f)$ in k space; i.e., in the spatial domain. In this case, it is natural to apply Taylor's frozen turbulence hypothesis, and thus, wavenumber scales in the spatial domain can be replaced by frequency scales in the time domain; i.e., $k = 2\pi f/U$, where U is the mean wind velocity (in m/s) and f is frequency (in Hz). The spectra remain unchanged in shape as well as magnitudes. Thus, when interpreting lidar sounding data, it becomes important to know the profile of mean wind velocity in the atmosphere.

In this analysis, the single-scattering approximation is assumed. The influence of multiply scattered photons on the backscattered signal of a ground-based lidar system and on the aerosol extinction profiles retrieved from these signals was studied in the past by means of measurements and Monte Carlo simulations [63]. It was mentioned that for typical atmospheric optical depths ($\tau[500 \text{ nm}] \sim 0.15 \div 0.18$ optical depth) the multiple-scattering contribution to the backscattered signal does not exceed 5%. Therefore, in further considerations, the description can be limited by the single-scattering approximation.

Measurements in the atmosphere may not always satisfy an assumption of stationarity, or sampling rate may not be fast enough to avoid serious aliasing in the calculated spectrum. Some of the preprocessing techniques most commonly used minimize these effects. The data preprocessing includes trend removal and low-pass filtering procedures as well as the use of a tapering window [64, 65]. The backscattering coefficient fluctuation spectrum for different altitude elevations is then calculated by using the fast Fourier transform (FFT). From the spectrum, a value of the power law exponent α is calculated.

The trend can be defined as any frequency component with a period longer than the data sampling time, T. Sources of the trend may be diurnal variations in meteorological parameters, as well as transmitter and/or receiving sensor instability. The presence of a trend in the time series makes the data nonstationary and produces distortions in the low-frequency tail of the spectrum. This distortion can mask the real behavior of the spectrum on these scales.

The procedure of trend removal includes a low-order polynomial fitting to the data (the order of polynomial should be ≤4) and subtracting the fitted values from the original time series [64]. A third-order polynomial fitting can be used in the data processing.

The data sampling time, T, establishes the lowest frequency $(1/T)$ that can be resolved as well as the width of each elementary frequency band, Δf, in the spectrum: $\Delta f = 1/T = 1/p\Delta t$, where p is the number of samples. Data sampling rate Δt determines the frequency bandwidth $(f = 1/\Delta t)$ in the calculated spectra. Because the spectral information at frequencies above $f_N = 1/[2\Delta t]$ (Nyquist frequency) is folded back into the spectrum, low-pass filtering is applied to reduce the energy above f_N.

A tapering window is applied to the time series to bring the values down to zero at both ends of the sampling period, which reduces the discontinuity at the boundaries of the data and the "side-band leakage" effect (the leakage of energy to neighboring frequencies through the side lobes in the convolving function). Thus, it minimizes the effect of finite sampling on the magnitude of the computed spectrum and improves the ability to resolve discrete contributions to the spectrum from waves in the signal. A low-pass second-order Butterworth filter and cosine-tapered Tukey's window can be applied.

The behavior of spectrum depends on the cutoff frequency, f_{cutoff}, in signal low-pass filtering. In general, the minimal aerosol inhomogeneity size (or aerosol-cloud size) that can be registered corresponds to the cross-sectional area of the volume illuminated by the laser beam at given altitude $d(h)$. In this case, the time t_p of aerosol inhomogeneity passage through the illuminated volume at a selected altitude is $t_p = d(h)/U(h)$, where $U(h)$ is the wind velocity at altitude h.

To register the aerosol inhomogeneities of the diameter, d, at least two samples are needed per period. In this case, the upper limit for the sampling interval, t_s (or laser pulse repetition rate, f_s), needs to be

$$t_s \leq \frac{1}{2 \cdot f_p} \quad \text{or} \quad f_s \geq 2 \cdot f_p \tag{3.26}$$

where f_p is the signal bandwidth ($f_p = 1/t_p$). Let $d(h)$ be the diameter of the illuminated cross-section area at a selected altitude h viewed by a receiver with a given FOV, and Δt be the sampling interval (or sampling frequency, $\Delta f = 1/\Delta t$). The highest measurable frequency is determined as $f_N = 1/[2\Delta t]$.

Following Equation 3.26, the low-path filter cutoff frequency can be chosen as 5–10 times smaller than f_N. At a given laser pulse rate Δt, we can write for the minimal aerosol inhomogeneity size that can be registered, $d_N(h)$: $d_N(h) = U(h)/f_N$. Thus, a low-pass filter cutoff frequency should be selected as

$$f_{\text{cutoff}} = \begin{cases} \dfrac{U(h)}{\alpha_1 \cdot d(h)}, & \text{if} \quad d_N(h) < d(h) \\[2ex] \dfrac{U(h)}{\alpha_2 \cdot d_N(h)}, & \text{if} \quad d_N(h) \geq d(h) \end{cases} \tag{3.27}$$

where $\alpha_1 = 5$ and $\alpha_2 = 10$ can be used, and the minimal aerosol inhomogeneity size, which can be retrieved from measurements, corresponds to $L_{\min} = U(h)/f_{\text{cutoff}}$.

The noise level of the measured signal can be estimated from the time series of backscattered intensity fluctuations taken at the maximal altitude, where the lidar return corresponds to background and electronic noise. The noise power spectral density (PSD_N) and noise variance (σ_N^2) are calculated by use of FFT, and the SNR for different altitudes can be estimated. The SNR is given as follows:

$$SNR = \frac{\sigma_S^2}{\sigma_N^2} \quad \text{or} \quad SNR = \frac{\sum_f PSD_S}{\sum_f PSD_N} \tag{3.28}$$

where σ_S^2 and PSD_S are the variance and power spectral density of backscattered signal, respectively.

PSD is the frequency response of a random or periodic signal. By definition, the PSD of a random time signal, $P(t)$, is the Fourier transform magnitude squared over a time interval, T. It can be expressed as

$$\text{PSD} \equiv W(f) = \frac{1}{T}\left|\int_{-T/2}^{T/2} P(t)\exp[-j2\pi ft]dt\right|^2 \tag{3.29}$$

As mentioned previously, lidar measurements of the returned signal power fluctuations are performed in time. To have information about spatial modes at a given frequency f, it is necessary to know the mean velocity of inhomogeneities (in k space, where $k = 2\pi f/U$ and U is the mean wind velocity). It is noteworthy that the choice of wind speed model in the troposphere for the analysis of experimental data is a separate problem. Ideally, the lidar measurements should be accompanied by the wind velocity measurements.

Figure 3.12 shows an example of a normalized power spectrum, $W(z,f)/\sigma_S^2$, of the intensity fluctuations of a lidar signal from 4.07 km

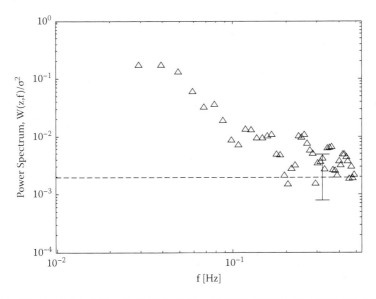

Figure 3.12 Normalized power spectrum, $W(z, f)/\sigma_S^2$, of the intensity fluctuations of a lidar signal over Beer-Sheva (Israel) at 4.07 km altitude elevation, 30 August 2004, 9:00 p.m. (local time); $\lambda = 532$ nm. (From A. Zilberman, E. Golbraikh, N. S. Kopeika, A. Virtser, I. Kupershmidt, and Y. Shtemler, "Lidar study of aerosol turbulence characteristics in the troposphere: Kolmogorov and non-Kolmogorov turbulence," *Atmos. Res.*, vol. 88, pp. 66–77, 2008. With Permission.)

altitude elevation. The dashed line corresponds to the mean noise power level with $\pm 1\sigma_N$ interval; SNR = 9.5. Figure 3.13 shows a frequency dependence of the reflected signal power for certain altitudes, obtained by processing experimental data and comparing it with a 1D power spectrum for the O-K case ($\alpha = 5/3$). The wind velocity profile for the given experimental conditions [142] is provided in Figure 3.14. In Figures 3.12 and 3.13, markedly different intervals for the frequencies $0.04 \leq f \leq 0.1$ Hz and $f > 0.1$ Hz are observed.

Figure 3.15 presents an altitude dependence of the spectral exponent α for the region $0.04 \leq f \leq 0.1$ Hz [66], which corresponds to the large spatial scale sizes. It can be seen that α increases with altitude, on the average, from 2.4 to 2.6. As shown in Reference 67, such behavior of α is connected with the fact that large-scale turbulence of the velocity field exerts a considerable influence on the passive scalar behavior.

A somewhat different behavior of the spectral exponent with altitude is observed within the interval $f > 0.1$ Hz. It is shown in Figure 1.7. Here, α also grows with altitude and varies from values close to 4/3–5/3 (helical-O-K spectrum [68, 69]) to 3 (independent anisotropic spectrum [67]). It is noteworthy that the spectrum with $\alpha = 3$ applied to optical studies was first investigated in References 70 and 71. However, in those papers, that spectrum was connected with the stratosphere and gravitation waves. Apparently, this spectrum manifests itself in the behavior of the passive scalar (in the present case, aerosol inhomogeneities) even in the troposphere, at altitudes starting from ~6 km.

The possibility of such studies was shown in References 60 and 62 for a different geographical zone and altitudes up to 5 km. As follows from experimental data obtained in References 60, 62, and 66, the spectrum of turbulent aerosol fluctuations is rather complicated and not universal. It means that it is described by different values of spectral exponent α in different frequency regions (see Figures 3.12 and 3.13).

It should be pointed out that although the experimental studies of the behavior of turbulent fluctuations in the troposphere refer to the Mediterranean zone, the methods of study of the behavior of aerosol turbulent fluctuation dependence on the altitude are applicable to any geographic zone.

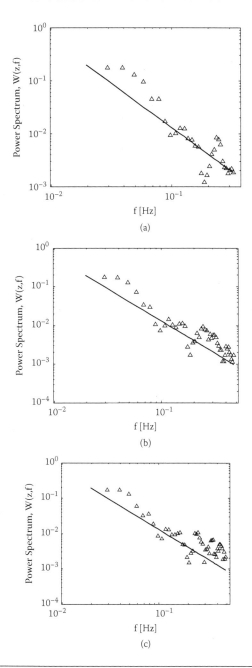

Figure 3.13 Power spectrum, $W(z, f)$, of the intensity fluctuations of a lidar signal for various altitudes, z: (a) −2.27 km; (b) −3.02 km; (c) −4.07 km; (d) −6.7 km. The solid line in each graph corresponds to a = 5/3. (From A. Zilberman, E. Golbraikh, N. S. Kopeika, A. Virtser, I. Kupershmidt, and Y. Shtemler, "Lidar study of aerosol turbulence characteristics in the troposphere: Kolmogorov and non-Kolmogorov turbulence," *Atmos. Res.*, vol. 88, pp. 66–77, 2008. With Permission.)

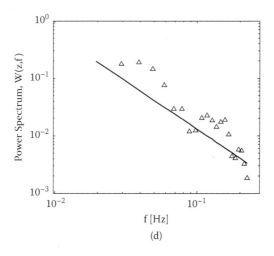

(d)

Figure 3.13 *(Continued)*

3.3 Lidar Measurement of Atmospheric Aerosol Parameters

Atmospheric aerosols that scatter and absorb electromagnetic radiation also reduce the performance of electro-optical systems operating through the atmosphere.

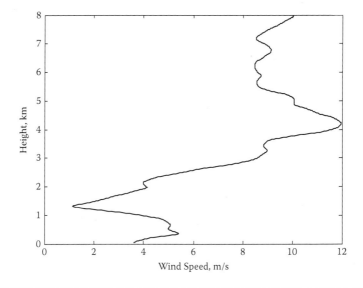

Figure 3.14 Wind velocity profile used in calculations. (From A. Zilberman, E. Golbraikh, N. S. Kopeika, A. Virtser, I. Kupershmidt, and Y. Shtemler, "Lidar study of aerosol turbulence characteristics in the troposphere: Kolmogorov and non-Kolmogorov turbulence," *Atmos. Res.*, vol. 88, pp. 66–77, 2008. With Permission.)

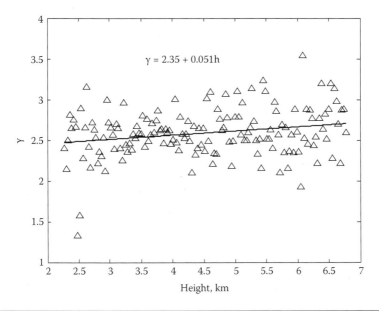

Figure 3.15   Changes in spectral exponent α with altitude for frequency interval of 0.04 Hz ≤ f ≤ 0.1 Hz; 30 m altitude resolution. (From A. Zilberman, E. Golbraikh, N. S. Kopeika, A. Virtser, I. Kupershmidt, and Y. Shtemler, "Lidar study of aerosol turbulence characteristics in the troposphere: Kolmogorov and non-Kolmogorov turbulence," *Atmos. Res.*, vol. 88, pp. 66–77, 2008. With Permission.)

The performance estimation of electro-optical systems depends on the accuracy of the atmospheric aerosol models being used in the propagation prediction codes. Large uncertainties remain in the models for aerosol constituents. Difficulties in estimating the influence of aerosols on electro-optical system performance and their climatic impact arise from the high spatial and temporal variability of aerosol concentrations, and physical and chemical properties. In addition, the large size range makes the measurements of aerosols very difficult, time consuming, and susceptible to error.

Lidar backscatter varies as a function of wavelength, as well as atmospheric composition and aerosol properties. The spectral distribution of the backscattering is strongly influenced by particle size distribution, so that by using lidar at more than one wavelength, one can obtain information about the aerosol size distribution (ASD) at different elevation heights.

Although knowledge of columnar ASD is obtained on a regular basis by inversion of spectral optical depths measured with multiwavelength radiometers [72, 73], information on the altitude-resolved ASD is rather limited, particularly in the troposphere, and is mainly based on airborne and balloon-borne measurements [74]. A significant database on stratospheric aerosol characteristics is, however, obtained by means of satellite-based spectral extinction measurements [75, 76]. Several attempts have been made to obtain information on the ASD by use of remote measurements of the vertical distribution of aerosol extinction and backscatter coefficients by multiwavelength lidars [77–82].

The aerosol volume extinction (or backscatter) coefficient at a given wavelength is the integral of the particle extinction (or backscatter) cross-section of different sizes weighted over the particle number size distribution. In Reference 77, the feasibility of determining the ASD from aerosol backscatter and extinction coefficients at four lidar wavelengths was numerically studied using a randomized minimization search technique [83]. It was found that accurately determined backscatter and extinction coefficients at four wavelengths carry enough information for retrieval of the ASD if the complex refractive index is known precisely. In Reference 78, the potential of a multiwavelength lidar for discriminating between several aerosol types, such as maritime, continental, stratospheric, and desert, was demonstrated on the basis of the wavelength dependence of the aerosol backscatter coefficient.

In Reference 79, a regularization method to infer ASD from volume extinction and backscatter coefficients observed by multiwavelength lidars was proposed, assuming that these parameters can be accurately calculated from the lidar data. In Reference 82, the mode radius and the corresponding half-width was retrieved by fitting the multiwavelength lidar data to a monomodal lognormal ASD. In Reference 84, two iterative methods of inverting lidar backscatter signals to determine altitude profiles of aerosol extinction and altitude-resolved ASD were presented.

Despite the large set of measurements, information about atmospheric aerosol optical properties remains fractional and limited, especially as a function of altitude. The lidar response may be interpreted in terms of the *lidar range equation* [1, 4]. The instantaneous received power, $P(z)$, due to backscattering from height z, assuming only *single elastic scattering* and a vertically pointing lidar, may be formally expressed in the form

$$P(z) = P_0 \cdot G(z) \cdot \xi(\lambda) \cdot \frac{ct_p}{2} \cdot \frac{A_R}{z^2} \cdot [\beta_\pi(z) + \beta_{\mathrm{mol}}(z)] \cdot T^2(z) \quad (3.30)$$

where

P_0 is the transmitted peak power (W),

$G(z)$ is the telescope-laser beam overlap function (G factor),

ct_p is the transmitted pulse length (m) (t_p is the pulse duration; c is the speed of light),

A_R is the receiver outer aperture area (m^2),

A_R/z^2 is the solid angle (sr) subtended by receiver aperture A_R at range z,

$\beta(z)$ is the backscatter coefficient at the aerosol volume location ($km^{-1}sr^{-1}$),

$\beta_{mol}(z)$ is the backscatter coefficient of air molecules,

$T^2(z)$ is the atmospheric transmission to range z and back to the receiver,

$\xi(\lambda)$ is the receiver spectral transmission factor, which is usually determined from the interference filter width and system optical losses.

The transmission or attenuation factor $T(\lambda, z)$ follows the exponential law of attenuation, the Beer-Lambert law. It is related to the integrated extinction by

$$T(\lambda, z) = \exp\left(-\int_0^z [\alpha(\lambda, z') + \alpha_{mol}(\lambda, z')]dz'\right) \qquad (3.31)$$

where $\alpha(\lambda, z)$ is the range-dependent aerosol extinction coefficient (km^{-1}) and $\alpha_{mol}(\lambda, z)$ is the extinction coefficient of air molecules. The extinction coefficient includes the loss effects of both scattering and absorption.

It is assumed that the emitted pulse width is much shorter than the acquisition period of the receiving instrumentation (acquisition card, photon counter, or simple amplitude scope). The pulse length determines the minimum spatial resolution of the lidar: $R_{min} = ct_p/2$.

Because the transmitted pulse has a finite duration, it illuminates a finite geometrical length, ct_p, of the atmosphere at any instant. However, because the received energy must travel a two-way path, the atmospheric length (or range increment) from which signals are received at any time is just half this value. Knowing the digitizer sample rate, it is possible to derive the spatial elevation resolution of the system.

For a lidar system with a narrow FOV and the separation between transmitter and receiver optical axes, an exact interpretation of lidar return signal in the short range is complicated by an unknown factor, called the geometrical form factor (*G* factor), of incomplete overlap between the transmitted laser beam and the FOV of the receiving optics. To interpret the lidar signal properly at short ranges, it must be corrected by the *G* factor.

To retrieve the optical parameters of interest (vertical profiles of the aerosol extinction and backscatter coefficient, optical depth), it is necessary to use an appropriate inversion algorithm of the lidar signals, because the lidar equation (3.30) contains two unknowns: the volume backscattering coefficient, β, and the extinction coefficient, α. Unless a cooperative Raman lidar channel is used, independent inversion of the optical parameters is not possible from a single backscatter channel. Consequently, a physical or a priori assumption to derive the extinction profile is introduced.

Various lidar inversion methods have been developed. Most of them use a single-wavelength algorithm, although there are certain algorithms using two or more wavelengths [9]. Collis's slope method [85, 86] estimates the atmospheric extinction and backscatter optical parameters under the assumption of a homogeneous atmosphere. To counteract the limitations of atmospheric homogeneity, other variants of the method exist: the slice method [87] and the two-angle method [88].

One widely used method is Klett's inversion method [89, 90]. This method is the most stable in practical applications. To retrieve the aerosol extinction profiles, a two-component lidar equation solution (Klett-Fernald method [89–91]) using backward integration and a noise-proof technique [92] can be utilized. Following Klett's method, the calculation of the extinction coefficient starts with the lidar range-corrected signal $X(z)$:

$$X(z) = \frac{P(z) \cdot z^2}{G(z)} = \beta(z) \cdot T^2(z) \qquad (3.32)$$

where $G(z)$ is the *G* factor. Taking the derivative of the logarithm of $X(z)$ and solving the differential equation (the Bernoulli equation), we have:

$$\alpha(z) = \frac{\exp[(S - S_m)/b]}{\alpha_m^{-1} + \frac{2}{b}\int_z^{z_m} \exp[(S - S_m)/b]dz'} \qquad (3.33)$$

where $S = \ln[X(z)]$, $S_m = \ln[X(z_m)]$, $\alpha_m = \alpha(z_m)$, and $z < z_m$. This is a so-called far-end solution of the lidar equation. The basic principle of this method is to start the inversion at the far end of the range interval (z_m). The method requires knowledge or at least an estimate of the extinction coefficient α_m at that point; but the influence of this boundary value on the resulting extinction coefficient profile decreases strongly with the optical thickness of the range interval considered.

The advantage of this method is that the system constant no longer needs to be determined, and fairly accurate retrievals of $\alpha(z)$ can be obtained for rather crude estimates of α_m. However, the assumption of a power law relationship between aerosol extinction and backscattering must be made. It is assumed that

$$\beta(z) = A\alpha^b \qquad (3.34)$$

where A and b are constants. As typically applied (for single-scatter conditions), b is set to unity, which is identical to assuming the aerosol extinction to backscatter ratio, $S_b(z)$ (called *lidar ratio*), is constant. If the lidar ratio is taken as a constant and not as dependent on altitude, then the extinction retrieving is similar to that obtained by Fernald [91]. This ratio, however, depends on the size distribution and the refractive index of the aerosol particles as well as on the lidar wavelength.

In Reference 93, the wavelength and humidity dependence of S_b was examined using Mie theory [94], based on observed ASD and refractive index data and obtained average values for S_b of 66 (17), 60 (13), 52 (13), and 42 (11) for wavelengths of 355, 532, 694, and 1064 nm, respectively. The standard deviations of the values (indicated in parentheses) are significant. In Reference 95, using simultaneous measurements by lidar, sun photometer, and optical particle counter, the value of S_b was estimated to range from 20 to 70. These studies thus show that the value of S_b varies significantly with aerosol characteristics.

The lidar ratio can be used to locate regions of atmospheric aerosol layering and to provide information on the transmission and reflection properties of the atmosphere. It is a relative measure of ASDs with respect to height, with larger particles generally being represented by smaller lidar ratios. For instance, a change of the lidar ratio from 18 to 36 sr corresponds to a change of the effective radius from 0.35 to 0.18 μm. From the aerosol extinction profiles, the aerosol layer optical thickness, atmospheric transmission, atmospheric boundary layer height, and ASD can be determined.

The spectral distribution of the backscatter and extinction coefficients is strongly influenced by the particle size distribution (Equations 1.22 and 1.23), so that by using lidar at more than one wavelength, one can obtain information about the ASD. The function α_λ (Equation 1.22) is measured in the experiment (Equation 3.33), $\pi r^2 Q_{ext}$ is the kernel (which can be calculated using Wiscombe's improved Mie scattering algorithm [96]), and $N(r)$ is the unknown function.

The effective radius range that can be retrieved from the measured extinction coefficients at wavelengths of 355, 532, and 1064 nm (corresponding to a lidar system with Nd:YAG laser) is about from 0.08 to 1.2 μm, and covers the accumulation mode of atmospheric aerosol particles. This particle size range is very effective for scattering of radiation at ultraviolet (UV), visible, and infrared (IR) wavelengths. It should be noted that the effect of variations of the refractive index of particles on the distribution shape is small in the size range mentioned previously.

The main difficulty in retrieving the particle size distribution from measurements of light extinction at several wavelengths lies in the solution of the Fredholm integral equation of the first kind (Equation 1.22). In practice, this so-called ill-posed problem invariably leads to a highly unstable solution because even arbitrarily small noise components in the measured quantities can give rise to extremely large spurious oscillations in the solution.

The ASD for different altitudes can be retrieved from measured extinction profiles using the suitable inversion method [48–56, 97]. Some examples of particle size distributions at different altitudes as well as lidar-derived vertical profiles of particle concentrations (number and volume) are shown in Figures 3.16–3.18. Figure 3.16 shows a vertical profile of number (a) and volume (b) concentration derived from the lidar data on 14 June 2004 (Beer-Sheva, Israel) [42, 59]. In Figure 3.17, number concentrations as a function of altitude retrieved from lidar measurements on 21 June 2004 are shown and compared with the AFGL "clear" and "hazy" atmospheric model [98, 99].

Profiles show increased aerosol extinction in the mid-troposphere and relatively thick layers at altitudes between 1 and 6 km. These are somewhat typical for desert dust and are related to the dust

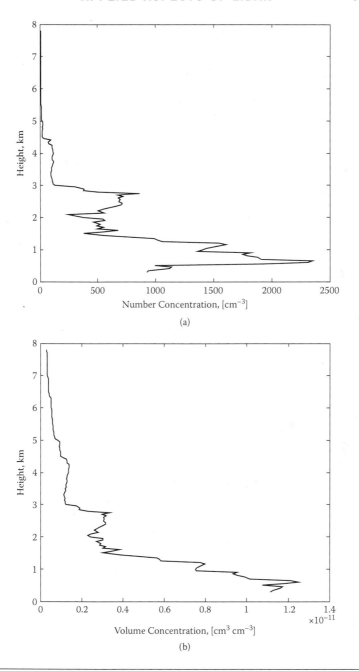

Figure 3.16 Number (a) and volume (b) concentration versus. altitude retrieved from lidar measurements on 14 June 2004 in Beer-Sheva (Israel) at 9:30 p.m. (local time) (From A. Zilberman and N. S. Kopeika, "Aerosol and turbulence characterization at different heights in semi-arid region," in *Atmospheric Optical Modeling, Measurement, and Simulation, Proc. SPIE*, no. 5891, pp. 129–140, 2005. With Permission.)

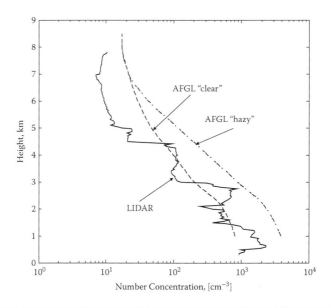

Figure 3.17 Number concentration versus altitude (solid) retrieved from lidar measurements on 21 June 2004 in Beer-Sheva (Israel) at 9:30 p.m. (local time) and compared with the AFGL "clear" (dot) and "hazy" (dash-dot) atmospheric model. (From A. Zilberman and N. S. Kopeika, "Aerosol and turbulence characterization at different heights in semi-arid region," in *Atmospheric Optical Modeling, Measurement, and Simulation, Proc. SPIE*, no. 5891, pp. 129–140, 2005. With Permission.)

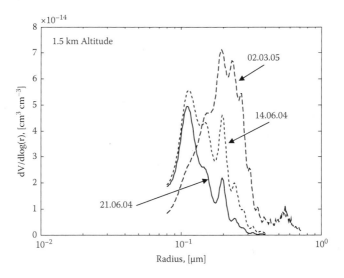

Figure 3.18 Lidar-measured aerosol volume size distribution at 1.5 km altitude for different days: 2 March 2005 (long dash); 14 June 2004 (short dash); 21 June 2004 (solid). (From A. Zilberman and N. S. Kopeika, "Aerosol and turbulence characterization at different heights in semi-arid region," in *Atmospheric Optical Modeling, Measurement, and Simulation, Proc. SPIE*, no. 5891, pp. 129–140, 2005. With Permission.)

mobilization process and to the strong convective activity occurring over desert areas, mainly at low latitudes.

Figure 3.18 shows an example of the lidar-derived aerosol volume concentration as a function of particle size for different days in March 2005 and June 2004. The results of long-time observations over different places show that tropospheric aerosol layers appeared around temperature inversion at heights 1.5–3 km, 5–7 km, and mostly in the tropopause region, ~9–10 km. As mentioned previously, there is a boundary between near-surface aerosols and tropospheric aerosols at heights about 0.5–1 km. Above this boundary, a reduction of number concentration of accumulation mode fraction as well as coarse mode can be observed. However, the concentration can increase if a temperature inversion at the boundary of a mixing turbulent layer (1.5–2.5 km) exists.

Measurements near ground (surface layer, up to 30 m) show that there is correlation between changes in turbulence strength and aerosol concentration values [100]. It was observed that increased aerosol concentration leads to increases in turbulence strength. Depending on wind speed, this may be a result of additional warming of ambient air in aerosol layers due to the increased aerosol absorption of radiation.

The results of lognormal-fitted average size distributions retrieved from lidar measurements are given in Figure 3.19, which shows different models for comparison. In Figure 3.19, "Sharav" is the name of the local air stream (or local wind) that appears from the Sahara along Africa's beach across Egypt and Sinai. It is characterized by a specific distribution of atmospheric pressure, fast change in wind direction, sharp rise in temperature (sometimes 12–15°C in 10–12 h; i.e., more than 1°C·h^{-1}), and quick fall in relative humidity (up to 10–20%) [101]. It can involve winds with a surface velocity of ~30–50 km/h and dust storms.

Real-world ASDs involve aerosol transport by winds over large distances, so that actual aerosol populations often derive from different sources. This makes it difficult for models to correlate well with a high degree of predictability because different aerosol types such as marine, continental, desert, etc., may influence simultaneously the measurements in a given location.

At present, only sparse lidar data about vertical distribution of aerosols are available. These data point out some important features of aerosol/dust vertical profiles. For example, Hamonou et al. [102] analyzed dust plumes transported to the Mediterranean by two lidar

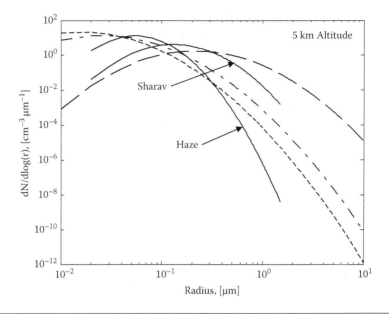

Figure 3.19 Lidar-derived average ASD at 5 km altitude (solid lines) for haze and Sharav condition compared to models: MODTRAN, background tropospheric (dash-dot); AFGL (dot); mineral transported, MODTRAN (dash) [42].

systems in different parts of the Mediterranean. According to these researchers, the transported dust over the Mediterranean is concentrated mainly at altitudes between 1.5 and 5 km. The observed dust transport is multilayered; several well-defined layers were detected in the free troposphere.

In addition, the observed aerosol layers can be connected to atmospheric turbulence layers or to changes in the turbulent behavior of air flows. Knowledge about particulate vertical distribution can provide additional information relating to the atmospheric turbulence field (altitude, strength, spectral behavior), and vice versa.

References

1. R. M. Measures, *Laser Remote Sensing: Fundamentals and Applications*, Wiley-Interscience, New York, 1984.
2. R. T. H. Collis, "Lidar," *Appl. Opt.*, vol. 9, pp. 1782–1788, 1970.
3. P. C. S. Devara, "Remote sensing of atmospheric aerosols from active and passive optical techniques," *Int. J. Remote Sensing*, vol. 19, no. 17, pp. 3271–3288, 1998.

4. E. D. Hinkley, Ed., *Laser Monitoring of the Atmosphere*, Springer-Verlag, Berlin, 1976.

5. Y. Y. Gu, C. S. Gardner, P. A. Castleberg, G. C. Papen, and M. C. Kelley, "Validation of the lidar in-space technology experiment: Stratospheric temperature and aerosol measurements," *Appl. Opt.*, vol. 36, no. 21, pp. 5148–5157, 1997.

6. J. A. Reagan, M. P. McCormick, and J. D. Spinhirne, "Lidar sensing of aerosols and clouds in the troposphere and stratosphere," *Proc. IEEE*, vol. 77, no. 3, pp. 433–448, 1989.

7. J. D. Spinhirne, "Micro pulse lidar," *IEEE Trans. Geosci. Remote Sensing*, vol. 31, no. 1, pp. 48–55, 1993.

8. M. J. McGill, "Lidar–remote sensing," in *Encyclopedia of Optical Engineering*, Marcel Dekker, New York, pp. 1103–1113, 2003.

9. V. A. Kovalev and W. E. Eichinger, *Elastic Lidar: Theory, Practice, and Analysis Methods*, John Wiley & Sons, New York, 2004.

10. P. S. Argall and R. J. Sica, "Lidar," in *Encyclopedia of Imaging Science and Technology*, J. P. Hornak, Ed., Wiley, New York, pp. 869–889, 2002.

11. M. Subramanian, "Laser remote sensing of atmospheric refractive-index-fluctuation profile," *J. Opt. Soc. Am.*, vol. 62, pp. 677–681, 1972.

12. R. R. Beland and J. Krause-Polstorff, *Lidar Measurements of Optical Turbulence: Theory of the Crossed Path Technique*, PL-TR-91-2139, Environmental Research Papers, no. 1089, Phillips Lab, Air Force Systems Comm., Hanscom AFB, Bedford, MA, 1991.

13. M. S. Belen'kii, "Effect of residual turbulent scintillation and a remote-sensing technique for simultaneous determination of turbulence and scattering parameters of the atmosphere," *J. Opt. Soc. Am.*, vol. 11, pp. 1150–1158, 1994.

14. M. S. Belen'kii and G. G. Gimmestad, "Design considerations for a residual turbulent scintillation (RTS) lidar," in *Atmospheric Propagation and Remote Sensing III*, W. A. Flood and W. B. Miller, Eds., *Proc. SPIE*, no. 2222, pp. 628–632, 1994.

15. M. S. Belen'kii and V. L. Mironov, "Laser method of determining the turbulence parameter C_n^2 on the basis of light scattering by atmospheric aerosol," *Radiophys. Quantum Electron.*, vol. 24, pp. 206–209, 1981.

16. M. S. Belen'kii, A. A. Makarov, V. L. Mironov, and V. V. Pokasov, "Lidar measurements of the structural characteristic of atmospheric turbulence," *Izv. Atmos. Oceanic Phys.*, vol. 20, pp. 328–330, 1984.

17. R. L. Schwiesov, "Effects of C_n^2 on a vertically pointing diffraction-limited lidar," *Appl. Opt.*, vol. 27, pp. 2517–2523, 1988.

18. M. S. Belen'kii and G. G. Gimmestad, "Monostatic image distortion technique for measuring intensity of atmospheric turbulence," in *Atmospheric Propagation and Remote Sensing III*, W. A. Flood and W. B. Miller, Eds., *Proc. SPIE*, no. 2222, pp. 621–627, 1994.

19. G. G. Gimmestad, J. R. White, and M. S. Belen'kii, "Design of an optical remote sensing system for measuring refractive turbulence in the Antarctic boundary layer," in *Image Propagation through the Atmosphere*, J. C. Dainty and L. R. Bissonnette, Eds., *Proc. SPIE*, no. 2828, pp. 221–231, 1996.

20. Y. A. Kravtsov and A. I. Saichev, "Properties of coherent waves reflected in a turbulent medium," *J. Opt. Soc. Am.*, vol. A2, pp. 2100–2105, 1985.

21. M. S. Belen'kii and G. G. Gimmestad, "A new remote sensing technique for lidar monitoring of atmospheric turbulence," in *Atmospheric Propagation and Remote Sensing II*, A. Kohnle and W.B. Miller, Eds., *Proc. SPIE*, no. 1968, pp. 596–606, 1993.

22. M. S. Belen'kii, D. W. Roberts, J. M. Stewart, G. G. Gimmestad, and W. R. Dagle, "Experimental validation of the differential image motion lidar concept," *Opt. Lett.*, vol. 25, pp. 518–520, 2000.

23. G. G. Gimmestad, M. W. Dawsey, D. W. Roberts, J. M. Stewart, J. W. Wood, F. D. Eaton, M. L. Jensen, and R. J. Welch, "Field validation of optical turbulence lidar technique," in *Atmospheric Propagation II*, C. Y. Young and G. C. Gilbreath, Eds., *Proc. SPIE*, no. 5793, pp. 10–16, 2005.

24. G. G. Gimmestad and M. S. Belen'kii, "Prospects for laser remote sensing of C_n^2," in *Atmospheric Propagation and Remote Sensing IV*, J. C. Dainty, Ed., *Proc. SPIE*, no. 2471, pp. 482–486, 1995.

25. V. P. Lukin and B. V. Fortes, *Adaptive Beaming and Imaging in the Turbulent Atmosphere*, SPIE Press, Bellingham, WA, 2002.

26. M. S. Belen'kii, "Tilt angular anisoplanatism and a full-aperture tilt-measurement technique with a laser guide star," *Appl. Opt.*, vol. 39, pp. 6097–6108, 2000.

27. V. P. Lukin, "Monostatic and bistatic schemes and an optimal algorithm for tilt correction in ground-based adaptive telescopes," *Appl. Opt.*, vol. 37, pp. 4634–4644, 1998.

28. V. P. Lukin, "Comparing the efficiencies of different schemes for laser guide stars forming," in *Adaptive Optics and Applications*, R. K. Tyson and R. Q. Fugate, Eds., *Proc. SPIE*, no. 3126, pp. 460–466, 1997.

29. R. Ragazzoni, "Robust tilt determination from laser guide stars using a combination of different techniques," *Astron. Astrophys.*, vol. 319, pp. L9–L12, 1997.

30. C. R. Neyman, "Focus anisoplanatism: A limit to the determination of tip-tilt with laser guide stars," *Opt. Lett.*, vol. 21, pp. 1806–1808, 1996.

31. F. Rigaut and E. Gendron, "Laser guide star in adaptive optics: The tilt determination problem," *Astron. Astrophys.*, vol. 261, pp. 677–684, 1992.

32. V. P. Lukin, B. V. Fortes, and E. V. Nosov, "Efficiencies of different schemes for the formation of laser guide stars," in *Adaptive Optical System Technologies*, D. Bonaccini and R. K. Tyson, Eds., *Proc. SPIE*, no. 3353, pp. 1109–1120, 1998.

33. M. S. Belen'kii, "Fundamental limitation in adaptive optics: How to eliminate it? A full aperture tilt measurement technique with a laser guide star," in *Adaptive Optics in Astronomy*, M. A. Ealey and F. Merkle, Eds., *Proc. SPIE*, no. 2201, pp. 321–323, 1994.

34. A. I. Kon, V. L. Mironov, and V. V. Nosov, "Fluctuations of the centers of gravity of light beams in a turbulent atmosphere," *Izv. Vuz. Radiofizika*, vol. 17, pp. 1501–1511, 1974.

35. V. L. Mironov and V. V. Nosov, "Random displacements of image in the focus of a telescope during lidar measurement of the turbulent atmosphere," *Izv. Vuz. Radiofizika*, vol. 20, pp. 1530–1533, 1977.

36. J. H. Churnside and R. J. Lataitis, "Angle-of-arrival fluctuations of a reflected beam in atmospheric turbulence," *J. Opt. Soc. Am.*, vol. A4, pp. 1264–1272, 1987.

37. J. H. Churnside, "Angle-of-arrival fluctuations of retroreflected light in the turbulent atmosphere," *J. Opt. Soc. Am.*, vol. A6, pp. 275–279, 1989.

38. V. M. Orlov, I. V. Samokhvalov, G. G. Matvienko, M. L. Belov, and A. N. Kozhemyakov, *The Elements of Theory of Wave Scattering and Optical Ranging*, Nauka, Novosibirsk, 1982 (in Russian).

39. V. L. Mironov, *Laser Beam Propagation in the Turbulent Atmosphere*, Nauka, Novosibirsk, 1981 (in Russian).

40. V. I. Tatarskii, *The Effects of the Turbulent Atmosphere on Wave Propagation*, Translated from the Russian by the Israel Program for Scientific Translations, Jerusalem, 1971.

41. A. Zilberman and N. S. Kopeika, "Laser beam wander in the atmosphere: Implications for optical turbulence vertical profile sensing with imaging LIDAR," *J. Appl. Remote Sensing*, vol. 2, 023540, 2008.

42. A. Zilberman, *Lidar Measurements of Turbulence Strength and Aerosol Profiles in the Negev Desert*, PhD Diss., Ben-Gurion University of the Negev, Beer-Sheva, Israel, 2007.

43. C. S. Cardner, B. M. Welsh, and L. A. Thompson, "Design and performance analysis of adaptive optical telescopes using laser guide stars," *Proc. IEEE*, vol. 78, pp. 1721–1743, 1990.

44. J.-L. Nieto and E. Thouvenot, "Recentering and selection of short-exposure images with photon-counting detectors," *Astron. Astrophys.*, vol. 241, pp. 663–672, 1991.

45. G. Cao and X. Yu, "Accuracy analysis of a Hartmann-Shack wavefront sensor operated with a faint object," *Opt. Eng.*, vol. 33, pp. 2331–2334, 1994.

46. J. Ares and J. Arines, "Influence of thresholding on centroid statistics: Full analytical description," *Appl. Opt.*, vol. 43, pp. 5796–5805, 2004.

47. A. Patwardhan, "Subpixel position measurement using 1D, 2D and 3D centroid algorithms with emphasis on applications in confocal microscopy," *J. Microscopy*, vol. 186, pp. 246–257, 1996.

48. D. L. Phillips, "A technique for the numerical solution of certain integral equations of the first kind," *J. Assoc. Comput. Mach.*, vol. 9, pp. 84–97, 1962.

49. A. N. Tikhonov, "On the solution of incorrectly stated problems and a method of regularization," *Dokl. Akad. Nauk. SSSR*, vol. 151, pp. 501–504, 1963.

50. J. M. Heneghan and A. Ishimaru, "Remote determination of the profiles of the atmospheric structure constant and wind velocity along a line-of-sight path by a statistical inversion procedure," *IEEE Trans. Antennas Propagat.*, vol. AP-22, pp. 457–464, 1974.

51. E. R. Westwater and A. Cohen, "Application of Backus-Gilbert inversion technique to determination of aerosol size distribution from optical scattering measurements," *Appl. Opt.*, vol. 12, pp. 1340–1348, 1973.

52. L. C. Chow and C. L. Tien, "Inversion techniques for determining the droplet size distribution in clouds: Numerical examination," *Appl. Opt.*, vol. 15, pp. 378–383, 1976.

53. S. Twomey, *Introduction to the Mathematics of Inversion in Remote Sensing and Indirect Measurements*, Elsevier, New York, 1977.

54. S. Twomey, "On the numerical solution of Fredholm integral equation of the first kind by the inversion of the linear system produced by quadrature," *J. Assoc. Comput. Mach.*, vol. 10, pp. 97–101, 1963.

55. M. D. King, "Sensitivity of constrained linear inversions to the selection of the Lagrange multiplier," *J. Atmos. Sci.*, vol. 39, pp. 1356–1369, 1982.

56. P. Qing, H. Nakane, Y. Sasano, and S. Kitamura, "Numerical simulation of the retrieval of aerosol size distribution from multiwavelength laser radar measurements," *Appl. Opt.*, vol. 28, pp. 5259–5265, 1989.

57. M. T. Chahine, "Determination of the temperature profile in an atmosphere from its outgoing radiance," *J. Opt. Soc. Am.*, vol. 58, pp. 1634–1637, 1968.

58. M. T. Chahine, "Inverse problems in radiative transfer: Determination of atmospheric parameters," *J. Atmos. Sci.*, vol. 27, pp. 960–967, 1970.

59. A. Zilberman and N. S. Kopeika, "Aerosol and turbulence characterization at different heights in semi-arid region," in *Atmospheric Optical Modeling, Measurement, and Simulation, Proc. SPIE*, no. 5891, pp. 129–140, 2005.

60. V. E. Zuev, B. D. Belan, and G. O. Zadde, *Optical Weather*, Novosibirsk, Nauka, 1990 (in Russian).

61. I. A. Razenkov and A. P. Rostov, "Lidar study of fluctuations spectrum of backscattering coefficient in the near ground atmospheric layer," *Atmos. Ocean Opt.*, vol. 6, pp. 1307–1316, 1993.

62. Yu. S. Balin, M. S. Belen'kii, V. L. Mironov, I. V. Samokhvalov, N. V. Safonova, and I. A. Resenkov, "Lidar investigations of the aerosol inhomogeneities in the atmosphere," *Izv. Atmos. and Ocean*, vol. 22, no. 10, pp. 1060–1064, 1986.

63. J. Ackerman, P. Volger, and M. Wiegner, "Significance of multiple scattering from tropospheric aerosols for ground-based backscatter lidar measurements," *Appl. Opt.*, vol. 38, pp. 5195–5201, 1999.

64. J. S. Bendat and A. G. Piersol, *Random Data: Analysis and Measurement Procedures,* John Wiley & Sons, New York, 1988.

65. F. J. Harris, "On the use of windows for harmonic analysis with the discrete Fourier transform," *Proc. IEEE*, vol. 66, pp. 66–67, 1978.

66. A. Zilberman, E. Golbraikh, N. S. Kopeika, A. Virtser, I. Kupershmidt, and Y. Shtemler, "Lidar study of aerosol turbulence characteristics in the troposphere: Kolmogorov and non-Kolmogorov turbulence," *Atmos. Res.*, vol. 88, pp. 66–77, 2008.

67. T. Elperin, N. Kleeorin, and I. Rogachevskii, "Isotropic and anisotropic spectra of passive scalar fluctuations in turbulent fluid flow," *Phys. Rev.*, vol. E53, pp. 3431–3441, 1996.

68. E. Golbraikh and N. Kopeika, "Behavior of structure function of refraction coefficients in different turbulent fields," *Appl. Opt.*, vol. 43, no. 33, pp. 6151–6156, 2004.

69. R. E. Good, B. J. Watkins, A. F. Quesada, J. H. Brown, and G. B. Loriot, "Radar and optical measurements of C_n^2," *Appl. Opt.*, vol. 21, pp. 3373–3376, 1982.

70. M. S. Belen'kii, S. J. Karis, J. M. Brown, and R. Q. Fugate, "Experimental study of the effect of non-Kolmogorov stratospheric turbulence on star image motion," *Proc. SPIE*, no. 3126, pp. 113–123, 1997.

71. A. S. Gurvich and M. S. Belen'kii, "Influence of stratospheric turbulence on infrared imaging," *J. Opt. Soc. Am.*, vol. A12, pp. 2517–2522, 1995.

72. G. E. Shaw, J. A. Reagan, and B. M. Herman, "Investigations of atmospheric extinction using direct solar radiation measurements made with multiple wavelength radiometer," *J. Appl. Meteorol.*, vol. 12, pp. 374–380, 1973.

73. K. K. Moorthy, P. R. Nair, and B. V. Krishna Murthy, "Size distribution of coastal aerosols: Effects of local sources and sinks," *J. Appl. Meteorol.*, vol. 30, pp. 844–852, 1991.

74. W. R. Leaitch and G. A. Isaac, "Tropospheric aerosol size distribution from 1982 to 1988 over eastern North America," *Atmos. Environ.*, vol. 25A, pp. 601–609, 1991.

75. J. Lenoble and C. Brogneiz, "Information on stratospheric aerosol characteristics contained in the SAGE satellite multiwavelength extinction measurements," *Appl. Opt.*, vol. 24, pp. 1054–1063, 1985.

76. P. H. Wang, M. P. McCormick, T. J. Swissler, M. T. Osborn, W. F. Fuller, and G. K. Yue, "Inference of stratospheric aerosol composition and size distribution from SAGE II satellite measurements," *J. Geophys. Res.*, vol. 94, pp. 8435–8446, 1989.

77. H. Muller and H. Quenzel, "Information content of multispectral lidar measurements with respect to the aerosol size distribution," *Appl. Opt.*, vol. 24, pp. 648–654, 1985.

78. Y. Sasano and E. V. Browell, "Light scattering characteristics of various aerosol types derived from multiple wavelength lidar observations," *Appl. Opt.*, vol. 28, pp. 1670–1679, 1989.

79. P. Quing, H. Nakane, Y. Sasano, and S. Kitamura, "Numerical simulation of the retrieval of aerosol size distribution from multiwavelength laser radar measurements," *Appl. Opt.*, vol. 28, pp. 5259–5265, 1989.

80. M. J. Post, C. J. Grund, A. O. Langford, and M. H. Proffitt, "Observations of Pinatubo ejecta over Boulder, Colorado by lidars of three different wavelengths," *Geophys. Res. Lett.*, vol. 19, pp. 195–198, 1992.

81. J. Kolenda, B. Mielke, P. Rairoux, B. Stein, M. Del Guasta, D. Weidauer, J. P. Wolf, L. Woste, F. Castagnoli, M. Morandi, V. M. Sacco, L. Stefanutti, V. Venturi, and L. Zuccagnoli, "Aerosol size distribution measurements

using a multispectral lidar system," in *Lidar for Remote Sensing*, R. J. Becherer, R. M. Hardesty, and J. P. Meyzonnette, Eds., *Proc. SPIE*, no. 1714, pp. 209–219, 1992.

82. M. D. Guasta, M. Morandi, L. Stefanutti, B. Stein, and J. P. Wolf, "Deviation of Mount Pinatubo stratospheric aerosol mean size distribution by means of a multiwavelength lidar," *Appl. Opt.*, vol. 33, pp. 5690–5697, 1994.

83. J. Heintzenberg, H. Muller, H. Quenzel, and E. Thomalla, "Information content of optical data with respect to aerosol properties: Numerical studies with a randomized minimization-search-technique inversion algorithm," *Appl. Opt.*, vol. 20, pp. 1308–1315, 1981.

84. K. Rajeev and K. Parameswaran, "Iterative method for the inversion of multiwavelength lidar signals to determine aerosol size distribution," *Appl. Opt.*, vol. 37, pp. 4690–4700, 1998.

85. R. T. H. Collis, "Lidar: A new atmospheric probe," *Q.J.R. Meteorol. Soc.*, vol. 92, pp. 220–230, 1966.

86. G. L. Kunz and G. De Leeuw, "Inversion of lidar signals with the slope method," *Appl. Opt.*, vol. 32, no. 18, pp. 3249–3256, 1993.

87. R. T. Brown, "New lidar for meteorological application," *J. Appl. Meteorol.*, vol. 12, pp. 698–708, 1973.

88. P. M. Hamilton, "Lidar measurements of backscatter and attenuation of atmospheric aerosols," *Atmos. Environ.*, vol. 3, pp. 221–223, 1969.

89. J. D. Klett, "Stable analytical inversion solution for processing lidar returns," *Appl. Opt.*, vol. 20, pp. 211–220, 1981.

90. J. D. Klett, "Lidar inversion with variable backscatter to extinction ratios," *Appl. Opt.*, vol. 24, pp. 1638–1643, 1985.

91. F. G. Fernald, "Analysis of atmospheric lidar observations: Some comments," *Appl. Opt.*, vol. 23, pp. 652–653, 1984.

92. Y. S. Balin, S. I. Kavkyanov, G. M. Krekov, and I. A. Razenkov, "Noise-proof inversion of lidar equation," *Opt. Lett.*, vol. 12, no. 1, pp. 13–15, 1987.

93. T. Takamura and Y. Sasano, "Ratio of aerosol backscatter to extinction coefficients as determined from angular scattering measurements for use in atmospheric lidar applications," *Opt. Quantum Electron.*, vol. 19, pp. 293–302, 1987.

94. H. C. van de Hulst, *Light Scattering by Small Particles*, Wiley, New York, 1957.

95. T. Takamura, Y. Sasano, and T. Hayasaka, "Tropospheric aerosol optical properties derived from lidar, sun photometer, and optical particle counter measurements," *Appl. Opt.*, vol. 33, pp. 7132–7140, 1994.

96. W. J. Wiscombe, "Improved Mie scattering algorithms," *Appl. Opt.*, vol. 19, pp. 1505–1509, 1980.

97. F. Ferri, A. Bassini, and E. Paganini, "Modified version of the Chahine algorithm to invert spectral extinction data for particle sizing," *Appl. Opt.*, vol. 34, no. 25, pp. 5829–5839, 1995.

98. E. P. Shettle and R. W. Fenn, *Models for the Aerosols of the Lower Atmosphere and the Effects of Humidity Variations on their Optical Properties*, AFGL-TR-79-0214, 1979.

99. R. A. McClatchey, R. W. Fenn, J. E. A. Selby, F. E. Volz, and J. S. Garing, *Optical Properties of the Atmosphere*, Air Force Cambridge Research Lab, Hanscom AFB, Bedford, MA, AFCRL-72-0497, 1972.

100. N. S. Kopeika, *A System Engineering Approach to Imaging*, SPIE Press, Bellingham, WA, 1998.

101. N. Yackerson and A. Zilberman, "On the variations in the electrical state under specific meteorological conditions in the ground atmospheric layer in semi-arid areas," *Sci. Total Environ.*, vol. 347, pp. 230–240, 2005.

102. E. Hamonou, P. Chazette, D. Balis, F. Dulac, X. Scheider, E. Galani, G. Ancellet, and A. Papayannos, "Characterization of the vertical structure of Saharan dust export to the Mediterranean basin," *J. Geophys. Res.*, vol. 104, pp. 22,257–22,270, 1999.

4

OPTICAL COMMUNICATION CHANNELS

4.1 Main Characteristics

Optical wireless communication is rapidly becoming a familiar part of modern life [1, 2]. Over the last decade, there has been a steady increase in the number of consumers requiring high-capacity links. In the past, these customers were pleased with tens of megabits per second; but nowadays near-gigabits-per-second links are required. Banks, universities, offices, companies, and government facilities all need communication services for stationary and mobile applications. (The emerging technology of mobile optical wireless communication is beyond the scope of this book and will not be discussed in this chapter.) High-data-rate communication applications range from next-generation Internet and support for cellular infrastructure to last-mile applications. Some of the requirements of these new applications could be met by fiber optics or millimeter-wave wireless links. Fiber optics has been distributed in many cities in close proximity to the backbone of the network. However, a massive effort is required to bridge the distance from the central switch to the client premises (this is termed the "last mile" problem), and many difficulties need to be overcome. In some cases it may not be possible or practical, or it may be too time consuming or costly to dig up main streets and lay down fibers. In such cases a wireless solution can bridge the gap. Millimeter-wave wireless links provide medium data capacity for long ranges. However, the capacity is limited, and in some cases public health and safety considerations as well as heavy tariffs and licensing fees make this selection less favorable. Additionally, the bureaucracy involved in obtaining permits can take months. As a result, in cases where high data rate is required without any licensing and tariffs and the range is limited, optical wireless communication (OWC) is the best solution (Figure 4.1).

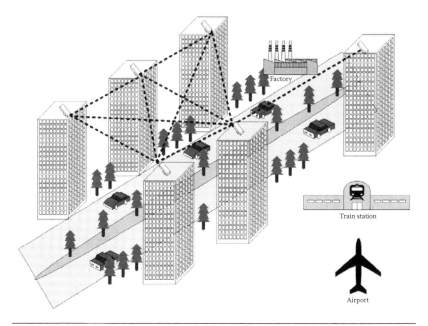

Figure 4.1   A schematic illustration of an optical wireless communication network. (From D. Kedar and S. Arnon, "Urban optical wireless communication network: The main challenges and possible solutions," *IEEE Commun. Magazine*, vol. 42, pp. S2–S7, 2004. With Permission.)

In addition, OWC is an excellent choice when a temporary solution is sought, or when an unexpected redeployment of premises calls for the provision of instant communication links. It is, of course, very important to design the system to operate in the eye-safe regime, in the interests of health and safety. After all, light in the infrared (IR) range has been harmlessly endured since the beginning of human evolution. In some scenarios a hybrid system is the optimum solution. A hybrid system includes an optical wireless transceiver, a millimeter radiofrequency (RF) wireless transceiver, a monitoring system, and a switch. The advantages of hybrid systems are (a) all-weather operation and (b) very high data rate most of the time. The disadvantages are the complexity and cost of the hybrid system. Most of the time the system transmits high-data-rate information, but when the weather becomes unstable, with a lot of haze or fog or very strong turbulence, the system switches to a medium data rate and transmits the information by millimeter RF wireless transceiver. The ability to work in all weather and to differentiate the type and priority of the information makes the system very reliable due to the facility that in the medium-data-rate mode we transmit high-priority information such as video streaming, browsing, and voice calls, while mail and backup are delayed. Following the previous discussion, the question "why not utilize present electrical cable infrastructures, which can be used to bridge the last-mile gap?" remains open. Electrical cable communication infrastructure includes, for example, asynchronous digital subscriber line (ADSL), power line communication (PLC), and cable television (CATV) connectivity. However, the limited bandwidth of these technologies and high leasing fees make them an inferior solution in comparison to OWC. In many cities all over the world the flexibility of OWC systems (also termed free-space optics [FSO]) provides high-capacity connections to the fiber backbone to home users. OWC can provide fiber-like performance using a very small, low-weight transceiver. In addition, the installation can take only a few hours, without entailing any licensing or fees. The installation process requires only electrical and communication link connections to the two transceivers and simple alignment between them. Transceivers can be placed on rooftops, billboards, electrical pillars, bridges, lampposts, or inside offices near windows, and an OWC system can be operative within hours. As the importance and value of the information transferred in the system increases, the security features of OWC become more significant. The security features of OWC result from the narrow-beam divergence angle of the transmitter, which prevents spillover of signal energy in an unintended direction, which could be used to eavesdrop (small footprint). Moreover, the narrow field of view (FOV) of the receiver reduces the probability of interference or jamming.

The main challenges in this field lie in extending the research work by providing tailored solutions for signal degradation due to atmospheric effects such as turbulence and aerosol scattering. OWC is being heralded as a unique communication technology for the coming decade, although some challenges must still be overcome.

4.1.1 Block Diagram of the Communication System

An optical wireless communication system comprises three main blocks [1, 2]: (a) the transmitter, (b) the channel, and (c) the receiver (Figure 4.2). The input to the transmitter is an electronic signal, which carries the information, and the output of the transmitter is

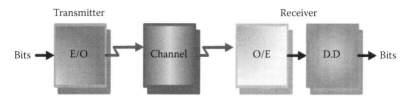

Figure 4.2 Optical wireless system scheme.

an optical signal from a light source, such as a light-emitting diode (LED) or laser. The optical signal carries the information. The input to the channel is the optical signal from the transmitter and the output of the channel is the input to the receiver. The receiver receives the optical signal from the output of the channel, amplifies the signal, converts it to an electronic signal, and extracts the information.

The transmitter includes a modulator, a driver, a light source, and optics (Figure 4.3). The modulator converts the information bits to an analog signal that represents a symbol stream. The driver provides the required current to the light source based on the analog signal from the output to the modulator. The light source is an LED or a laser, which is an incoherent or a coherent source, respectively. The source converts the electronic signal to an optical signal. The optics focuses and directs the light from the output of the source in the direction of the receiver.

The channel attenuates the power of the optical signal and widens and spreads it in the spatial, temporal, angular, and polarization domains (Figure 4.4). The attenuation, widening, and spreading are stochastic processes, resulting from interaction of light with atmospheric gases, aerosols, and turbulence. The atmospheric gases mostly absorb the light while the aerosols absorb and scatter the light. The

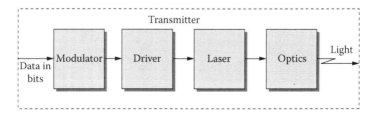

Figure 4.3 The transmitter scheme.

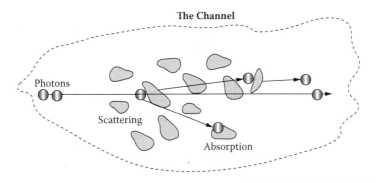

Figure 4.4   The channel. (From H. Manor and S. Arnon, "Performance of an optical wireless communication system as a function of wavelength," *Appl. Opt.*, vol. 42, pp. 4285–4294, 2003. With Permission.)

turbulence give rise to constructive and destructive interference, which cause fluctuations in received power, or scintillations, as well as spot wander in the detector plane, as described in Chapters 1 and 2.

The receiver includes optics, a filter, a polarizer, a detector, a transimpedance amplifier, a clock recovery unit, and a decision device (Figure 4.5). The optics concentrates the received signal power onto the filter. Only light at the required wavelength propagates through the filter to the polarizer. The polarizer only enables light at the required polarization to propagate through to the detector.

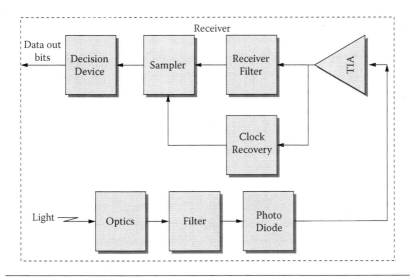

Figure 4.5   The receiver scheme. TIA = transimpedance amplifier.

The detector, in most cases a semiconductor device such as a positive intrinsic negative (PIN) photodiode, converts the optical signal to an electronic signal. The transimpedance amplifier amplifies the electronic signal from the detector. The clock recovery unit provides a synchronization signal to the decision device based on the signal from the output of the transimpedance amplifier. The decision device estimates the received information based on the electronic signal from the transimpedance amplifier and synchronization signal.

4.1.2 Link Budget

In this subsection, we describe the relation between the system parameters and the received power. We start with the definition of the concept of the gain of a telescope that we use in the following passage to describe the parameters of the transmitter and the receiver. It is clear that the telescope is a passive element, so the gain does not add any additional energy to the signal. The gain is the ratio of the radiation intensity of a telescope in a given direction to the intensity that would be produced by a telescope that radiates equally in all directions and has no losses. Therefore, the gain describes how the energy is distributed in the spatial domain.

The received power in the detector plane is given by [3]

$$P_R = P_T G_T T_T \left(\frac{\lambda}{4\pi Z} \right)^2 T_A G_R T_R T_F \tag{4.1}$$

where P_T is the transmitter optical power, G_T is the laser transmitter telescope gain, T_T is the optics efficiency of the transmitter, λ is the laser wavelength, and Z is the distance between the laser transmitter and the optical receiver. The term in parentheses is the free space loss, T_A is the atmospheric transmission, G_R is the optical receiver telescope gain, T_R is the optics efficiency of the receiver, and T_F is the filter transmission.

Figure 4.6 depicts the normalized received power as a function of distance. It is easy to see that as the distance increases, the power decreases to the power of two.

We use results from work done, and reported in Reference 3, to analyze the gain of the transmitter. In the analysis, the authors assume that a Cassegrain telescope is used and the laser emits a circular,

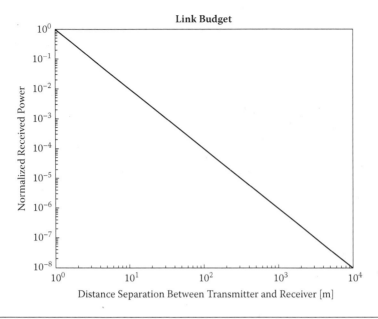

Figure 4.6 Normalized received power as a function of distance.

single-mode transverse electromagnetic (TEM) beam (Figure 4.7). Reference 3 gives the transmitter telescope gain, based on the previous assumption:

$$G_T(\alpha,\beta,\gamma,X) = \left(\frac{2\pi a}{\lambda}\right)^2 g_T(\alpha,\beta,\gamma,X), \qquad (4.2)$$

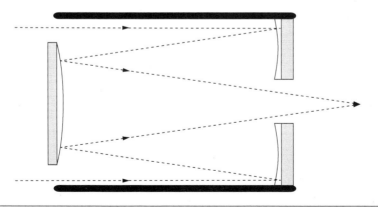

Figure 4.7 Cassegrain telescope.

where a is the radius of the aperture primary mirror and $g_T(\alpha,\beta,\gamma,X)$ is a function that takes into consideration the semi-obscuration, off-axis near-field effects, and defocusing effects. The arguments of function $g_T(\alpha,\beta,\gamma,X)$ are described in the next four equations:

$$\alpha = \frac{a}{\omega}, \tag{4.3}$$

$$\gamma = \frac{b}{a} \tag{4.4}$$

where b is the radius of the secondary mirror and ω is the distance from the principal axis to the e^{-2} intensity point of the laser radiation in the spatial domain. X is a parameter describing the off-axis distribution:

$$X = k\,a\,\sin(\theta_1) \tag{4.5}$$

β includes near-field and defocusing effects and is defined as follows:

$$\beta = \left(\frac{k\,a^2}{2}\right)\left[\frac{1}{\gamma} + \frac{1}{R}\right] \tag{4.6}$$

where θ_1 is the observation angle, $k = 2\pi/\lambda$, and R is the curvature of the phase front at the telescope aperture. The calculation of $g_T(\alpha,\beta,\gamma,X)$ requires integration of two exponential functions and one Bessel function of order zero and then taking the square of the absolute value of the integration result:

$$g_T(\alpha,\beta,\gamma,X) = 2\alpha^2 \left| \int_{\gamma^2}^{1} \exp(j\beta u)\exp(-\alpha^2 u) J_o[X(u)^{0.5}]du \right|^2 \tag{4.7}$$

To simplify Equation 4.7, we assume the special case of far-field, on-axis observation such that

$$g_T(\alpha,0,\gamma,0) = \frac{2}{\alpha^2}[\exp(-\alpha^2) - \exp(-\gamma^2\alpha^2)]^2 \tag{4.8}$$

It is possible to maximize the on-axis gain in the far field using the following relation:

$$\alpha \approx 1.12 - 1.3\gamma^2 + 2.12\gamma^4 \tag{4.9}$$

Equation 4.9 gives the maximum on-axis gain for the aperture-to-beamwidth ratio for a general obscuration and is accurate to within 41% for $y < 0.4$.

$$G_T = 10 \log \left(\frac{2\pi a}{\lambda} \right)^2 + 10 \log(g_T(\alpha, 0, \gamma, 0)) \quad \text{(dB)} \qquad (4.10)$$

In the analysis of the gain of the receiver telescope, similar assumptions are made; a Cassegrain telescope is used but, in this case, we assume that the receiver is placed far away from the transmitter so that plane waves impinge on the receiver aperture.

Reference 4 gives the receiver telescope gain, based on the previous assumption:

$$G_R = 10 \log \left(\frac{2\pi a}{\lambda} \right)^2 + 10 \log(1 - \gamma^2) + 10 \log \eta \quad \text{(dB)} \qquad (4.11)$$

where, for direct detection,

$$\eta = \frac{2}{1 - \gamma^2} \int_0^{\frac{R_D k}{2 F_s}} \frac{(J_1(u) - \gamma J_1(\gamma u))^2}{u} \, du \qquad (4.12)$$

Additionally, F_s is the F number of the telescope, defined as

$$F_S = \frac{f}{D} \qquad (4.13)$$

where f is the focal length of the telescope and D is the diameter of the entrance pupil.

Analysis of Equation 4.11 indicates that the first term is for an ideal unobscured telescope gain, the second term describes the loss due to blockage of the incoming light by the central obscuration, and the third term represents the losses in direct detection due to the spillover of signal energy beyond the detector boundary. Figure 4.8 depicts an ideal telescope gain as a function of wavelength. It is easy to see that one order of magnitude increase in wavelength reduces the gain by 20 dB.

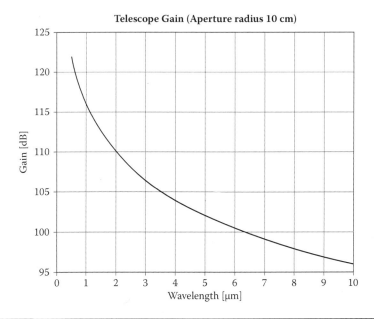

Figure 4.8 Ideal telescope gain as a function of wavelength.

4.2 Key Parameters Prediction

One of the important parameters that describes the atmosphere is power attenuation. During the last decades, many methods have been developed to calculate the attenuation based on meteorological data. One of the most popular methods takes advantage of visibility data that is regularly recorded at airports with very good temporal resolution (once every 30 minutes or less). The visibility is defined by several terminology sets that are very similar based on visual measurement or attenuation at 0.55 µm wavelength as a function of the range at which the image contrast drops to 0.02 and is given by [2]

$$V_S = \beta^{-1} \ln(1/0.02) = 3.912/\beta, \tag{4.14}$$

where β is the scattering coefficient and V_S is in kilometers.

From Equation 4.14 and References 5 and 6, the link attenuation is given by the following relation:

$$\tau = \exp\left[\frac{-3.91}{V_S}\left(\frac{\lambda}{0.55}\right)^{-q_S} Z\right], \tag{4.15}$$

where q_S is the size distribution parameter for scattering particles and Z is the propagation range in kilometers. q_S may typically be 1.6 for high visibility ($V_S > 50$ km), 1.3 for average visibility (6 km < V_S < 50 km), and $0.585\, V_S^{1/3}$ for low visibility ($V_S < 6$ km).

It is easy to show that γ, the attenuation coefficient of the link, is given by

$$\gamma = -\frac{1}{Z} Ln(\tau) \tag{4.16}$$

Attenuation is caused by atmospheric aerosol and molecular scattering and absorption. It can be expressed as

$$\gamma = \alpha_m + \beta_m + \alpha_a + \beta_a \tag{4.17}$$

where α and β represent absorption and scattering coefficients; subscript m refers to molecules and subscript a to aerosol particles.

The scattering and absorption coefficients are given by

$$\alpha = \sigma_a N_a \tag{4.18}$$

and

$$\beta = \sigma_s N_s \tag{4.19}$$

where σ represents cross-section parameters (m²) and N is particle concentration [1/m³]; subscript a refers to absorption and subscript s to scattering.

4.3 Mathematical and Statistical Description of Signal Fading

In many optical wireless communication systems, the received signal fades during typical operation conditions. The signal fade is a stochastic process caused by atmospheric turbulence. Turbulence is a phenomenon that describes random changes in the atmospheric refractive index in the spatial and temporal domains. This phenomenon is due to the temperature difference between the atmosphere, the ocean, and the ground and to the Earth's revolution. The stationary refractive index, n, of the atmosphere is a function of temperature, pressure, wavelength, and humidity (see Chapters 1 and 2). We briefly describe here some aspects that in our opinion can help the reader to understand the subject to be discussed. Thus, as was mentioned in Chapter 1, for a marine atmosphere one can present the refractive index in the following form [7]:

$$n_0 \approx 1 + \frac{77\,p}{T}\left[1 + \frac{7.53 \cdot 10^{-3}}{\lambda^2} - 7733\frac{q}{T}\right]10^{-6} \qquad (4.20)$$

where p is air pressure (millibars), T is temperature (K), and q is specific humidity (gm^{-3}).

To deal with the stochastic behavior of the refraction index, we use Kolmogorov's theory [7–10] close to the ground. One of the main parameters in this theory is C_n^2. C_n^2 is the refractive index structure constant that helps us to describe the fade statistics. Fried developed an analytic model to describe this constant [8], and later an improved model, named after its proposers Hufnagel and Stanley, became more acceptable [6, 12]. The latter model is given by (see also Chapter 1)

$$C_n^2(h) = 0.00594\left(\frac{v}{27}\right)^2 (10^{-5}\,h)^{10} \exp\left(-\frac{h}{1000}\right)$$

$$+2.7\ 10^{-16} \exp\left(-\frac{h}{1500}\right) + A\exp\left(-\frac{h}{100}\right) \qquad (4.21)$$

where h is the altitude, v is the wind speed, and A is the nominal value of $C_n^2(0)$ at zero altitude. At ground level, $C_n^2(0)$ typically ranges between 1.7×10^{-14} during daytime (strong turbulence) and 10^{-16} at night (weak turbulence).

When the channel is short or the turbulence is weak, the mathematical model for the covariance (over channel length L) for a plane wave in Kolmogorov turbulence is given by [7]

$$\sigma_X^2(L) = 0.56\left(\frac{2\pi}{\lambda}\right)^{\frac{7}{6}} \int_0^L C_n^2(u)(L-u)^{\frac{5}{6}}\,du. \qquad (4.22)$$

In addition, for a spherical wave it is given by

$$\sigma_X^2(L) = 0.56\left(\frac{2\pi}{\lambda}\right)^{\frac{7}{6}} \int_0^L C_n^2(u)\left(\frac{u}{L}\right)^{\frac{5}{6}} (L-u)^{\frac{5}{6}}\,du. \qquad (4.23)$$

The turbulence coherence diameter, d_0, in Kolmogorov turbulence is given for a plane wave by

$$d_0(L) = \left[1.45 \left(\frac{2\pi}{\lambda} \right)^2 \int_0^L C_n^2(u)du \right]^{-\frac{3}{5}}$$
(4.24)

and for a spherical wave by

$$d_0(L) = \left[1.45 \left(\frac{2\pi}{\lambda} \right)^2 \int_0^L \left(\frac{u}{L} \right)^{\frac{5}{3}} C_n^2(u)du \right]^{-\frac{3}{5}}$$
(4.25)

To analyze the temporal effect of the atmospheric turbulence, the frozen air model is used. In this model, it is assumed that the eddy pattern is stationary when it passes over the receiver plane; hence turbulence coherence time can be expressed as [11]

$$\tau_0 = \frac{d_0}{v},$$
(4.26)

where v is the wind velocity perpendicular to the beam propagation direction. In the following subsections, we describe the lognormal, the gamma-gamma, and the K-probability density distribution functions. The non-Kolmogorov turbulence channel is described in Chapters 1 and 2.

4.3.1 Lognormal Probability Density Function

The lognormal probability density function describes the scintillation and signal fading statistics for weak turbulence. The transmitted pulse propagates through a large number of elements of the atmosphere, each causing an independent, identically distributed (i.i.d.) phase delay and scattering. Use of the central limit theorem (CLT) indicates that the marginal distribution of the log-amplitude is Gaussian [11]:

$$f_X(X) = \frac{1}{\sqrt{2\pi\sigma_x^2}} \exp\left(-\frac{(X - E[X])^2}{2\sigma_x^2} \right)$$
(4.27)

where X is the log-amplitude fluctuation, σ_X^2 is the variance, and $E[X]$ is the ensemble average of log-amplitude X. In addition, X is assumed a homogeneous, isotropic, and independent Gaussian random variable.

The light intensity, I, as a function of X is given by

$$I = I_0 \exp(2X - E[X]) \tag{4.28}$$

where I_0 is the normalized signal light intensity.

The density distribution function of intensity, I, is lognormal [11]:

$$f_I(I) = \frac{1}{2I\sqrt{2\pi\sigma_x^2}} \exp\left(-\frac{(\ln(I) - \ln(I_0))^2}{8\sigma_x^2}\right), \tag{4.29}$$

In Figure 4.9, the lognormal density distribution function as a function of the intensity for $\sigma_x = 1$ and $\ln(I) = 3$ is shown. It can be seen that this function has one peak near the origin of the axes and then a sharp drop for increasing I.

4.3.2 Gamma-Gamma Density Distribution Function

The gamma-gamma distribution describes well a wide range of turbulence conditions from weak to strong. The gamma distribution is a

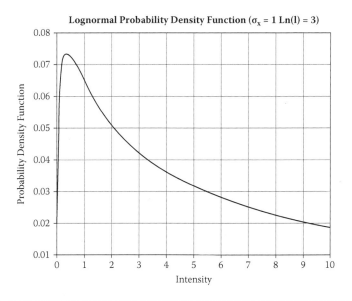

Figure 4.9   Lognormal density distribution function as a function of the intensity.

family of curves based on two parameters. The chi-square and exponential distributions are derived from the gamma-gamma distribution when one of the two gamma parameters is fixed. The gamma-gamma probability density function (PDF) is given by [12, 13]

$$f_I(I) = \frac{2(\alpha_t \beta_t)^{\frac{\alpha_t + \beta_t}{2}}}{\Gamma(\alpha_t)\Gamma(\beta_t)} I^{\frac{\alpha_t + \beta_t}{2} - 1} K_{\alpha_t - \beta_t}\left(2\sqrt{\alpha_t \beta_t I}\right) \quad I > 0 \quad (4.30)$$

where $K_{\alpha_t - \beta_t}(x)$ is the modified Bessel function of the second kind of order $\alpha - \beta$. The α_t and β_t are the effective number of small-scale and large-scale eddies of the scattering environment and $\Gamma(x)$ is the gamma function. The parameters α_t and β_t directly relate to large-scale and small-scale of scintillations of the optical wave and $\Gamma(x)$ is the gamma function. The parameters α_t and β_t are given by [12, 13]

$$\alpha_t = \left[\exp\left(\frac{0.49x^2}{\left(1 + 0.18d^2 + 0.56x^{\frac{12}{5}}\right)^{\frac{7}{6}}}\right) - 1\right]^{-1} \quad (4.31)$$

$$\beta_t = \left[\exp\left(\frac{0.51x^2\left(1 + 0.69x^{\frac{12}{5}}\right)^{-\frac{5}{6}}}{\left(1 + 0.9d^2 + 0.62d^2 x^{\frac{12}{5}}\right)^{\frac{7}{6}}}\right) - 1\right]^{-1} \quad (4.32)$$

Here

$$x^2 = 0.5C_n^2 k^{\frac{7}{6}} L^{\frac{11}{6}} \quad (4.33)$$

$$d = \sqrt{\left(\frac{kD^2}{4L}\right)} \quad (4.34)$$

where D is the diameter of the receiver collecting lens aperture and k is the wavenumber.

The gamma function is given by

$$\Gamma(z) = \int_0^\infty t^{z-1} e^{-t} dt \quad (4.35)$$

The recursive relation of gamma function is

$$\Gamma(z + 1) = z\Gamma(z) \tag{4.36}$$

and, if z is a positive integer,

$$\Gamma(z) = (z - 1)! \tag{4.37}$$

4.3.3 K Probability Density Distribution Function

The first non-Gaussian field models to gain wide acceptance for a variety of applications under strong fluctuation conditions were the family of K distribution, providing excellent models for predicting amplitude or irradiance statistics in a variety of experiments involving radiation scattered by turbulent media. The K-PDF has been successfully used to model atmospheric turbulence deep into saturation, and it is given by [14]

$$f_I(I) = \frac{2}{\Gamma(\alpha_n)\eta^{\alpha_n+1}} \alpha_n^{\frac{\alpha_n+1}{2}} I^{\frac{\alpha_n-1}{2}} K_{\alpha_n-1}\left(\frac{2}{\eta}\sqrt{\alpha_n I}\right) \tag{4.38}$$

Here I denotes the optical signal intensity and

$$\alpha_n = \frac{2}{\sigma_{si}^2 - 1} \tag{4.39}$$

where σ_{si}^2 is the scintillation index defined as

$$\sigma_{si}^2 = \frac{E\{I^2\}}{E^2\{I\}} - 1 \tag{4.40}$$

4.4 Modulation Methods

The modulation methods commonly used in optical wireless systems are based mostly on direct detection. The reason this kind of technology is preferred is that direct detection systems are simpler in comparison to coherent detection. Coherent detection requires a laser as a local oscillator. The laser should be matched to the received beam in its phase, polarization, and intensity profile. Furthermore, the development of optical amplifiers improves the performance of direct detection systems so that coherent detection provides an advantage of only a few decibels. As a result, direct detection is the popular way to detect signals. To detect the signal, we develop an algorithm to help us make the best decision. We define R_i as the received signal and S_i as the transmitted signal, and we limit the analysis to the binary situation. The probability that S_i was sent given that R_i was received is $P(S_i \mid R_i)$.

We will decide "1" in the case that

$$P(S_{zero} \mid R_i) < P(S_{one} \mid R_i) \tag{4.41}$$

and "0" in the case that

$$P(S_{zero} \mid R_i) > P(S_{one} \mid R_i) \tag{4.42}$$

Therefore, the decision criteria is given by

$$k = \frac{P(S_{zero} \mid R_i)}{P(S_{one} \mid R_i)} \langle \rangle 1 \tag{4.43}$$

According to this criteria, we calculate the probability of S_i given R_i. However, S_i is unknown and we only measure R_i. Therefore, we cannot calculate this probability. To solve this issue, we use Bayes's theorem:

$$P(A \mid B) = \frac{P(B \mid A)P(A)}{P(B)} \qquad P(B) \neq 0 \tag{4.44}$$

From Equations 4.43 and 4.44,

$$k = \frac{P(S_{zero} \mid R_i)}{P(S_{one} \mid R_i)} = \frac{\frac{P(S_{one})P(R_i \mid S_{one})}{P(R_i)}}{\frac{P(S_{zero})P(R_i \mid S_{zero})}{P(R_i)}} \langle \rangle 1 \tag{4.45}$$

Equation 4.45 can be simplified to

$$k = \frac{P(S_{one})P(R_i \mid S_{one})}{P(S_{zero})P(R_i \mid S_{zero})} \langle \rangle 1 \tag{4.46}$$

For wireless communication systems, we assume that $P(S_{one}) = P(S_{zero})$. Otherwise, data compression is possible, which results in equal probability: $P(S_{one}) = P(S_{zero})$. In this case, we get the maximum likelihood estimator (MLE), which is given by

$$k = \frac{P(R_i \mid S_{one})}{P(R_i \mid S_{zero})} \langle \rangle 1 \tag{4.47}$$

In the following subsections, we will review the most practical direct detection modulation schemes.

4.4.1 On-Off Keying Modulation

On-off keying (OOK) is a modulation method that uses two power levels to represent logic zero or logic one. In most cases, the logic zero is represented by a low power level, and the logic one is represented by a high power level. Figure 4.10 depicts an OOK signal. We start with the basics of detection theory and later provide the mathematical derivation for the case of the turbulence channel.

The error probability is given by

$$P_e = P(one) \int_{-\infty}^{V_T} P(v \mid one)dv + P(zero) \int_{V_T}^{\infty} P(v \mid zero)dv \tag{4.48}$$

where $P(one)$ is the a priori probability to transmit one, $P(zero)$ is the a priori probability to transmit zero, $P(v \mid one)$ is the probability to get v given that the transmission was one, and $P(v \mid zero)$ is the probability to get v given that the transmission was zero. To find the value of

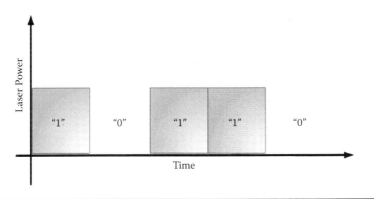

Figure 4.10 On-off keying modulation.

the threshold, V_T, that minimizes the error probability, we take the derivative of Equation 4.48 and compare it to zero, such as [15]

$$\frac{dP_e}{dV_T} = 0 \qquad (4.49)$$

Using the Leibniz integral rule, which gives a formula for differentiation of a definite integral whose limits are functions of the differential variable, we can write

$$P_e = P(one)P(V_T \mid one) - P(zero)P(V_T \mid zero) = 0 \qquad (4.50)$$

We also use the assumption that the a priori probabilities are equal, as follows:

$$P(one) = P(zero) = 0.5 \qquad (4.51)$$

From Equations 4.50 and 4.51, it follows that

$$P(V_T \mid one) = P(V_T \mid zero) \qquad (4.52)$$

The solution of Equation 4.52 is the intersection value that provides the V_T that minimizes the error probability. The graphic and literal representation of this problem is depicted in Figure 4.11 and described in the following sentences. In this figure, two Gaussian curves represent the PDF of zero and one.

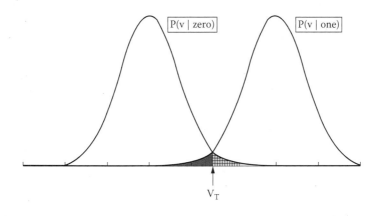

Figure 4.11   Calculation of error probability.

The main question in detection theory is to calculate the threshold that will minimize the probability of error. The probability of error is the sum, of the area below the curve of zero that is bound by the threshold and infinity and the area below the curve of one that is bound by minus infinity and the threshold. It is easy to see that for most cases the value of threshold that minimizes the error probability is the intersection of the two curves. If the threshold moves from the intersection to the right, we add additional area from the curve of one, and if the movement is to the left, we add additional area from the zero. Therefore, the intersection is the optimum point.

We will assume that either the background or the thermal noise is the dominant noise source, so the noise can be modeled by additive white Gaussian noise with zero mean and covariance σ_N^2 that is statistically independent of the received signal. Therefore, the probabilities $P(v \mid one)$ and $P(v \mid zero)$ are given by

$$P(v \mid one) = \frac{1}{\sqrt{2\pi\sigma_{one}^2}} e^{-\left(\frac{(v-\mu_1)^2}{2\sigma_1^2}\right)} \tag{4.53}$$

$$P(v \mid zero) = \frac{1}{\sqrt{2\pi\sigma_{zero}^2}} e^{-\left(\frac{(v-\mu_0)^2}{2\sigma_0^2}\right)} \tag{4.54}$$

In the case of symmetry, where $\sigma_1^2 = \sigma_0^2$, it is easy to see that the value of threshold that minimizes the error probability is given by

$$V_T = \frac{\mu_1 + \mu_0}{2} \tag{4.55}$$

$$P_e = \frac{1}{2} Q\left(\frac{\mu_1 - V_T}{\sigma_n}\right) + \frac{1}{2} Q\left(\frac{-\mu_0 + V_T}{\sigma_n}\right) \tag{4.56}$$

where

$$Q(x) = \frac{1}{\sqrt{2\pi}} \int_0^x e^{-\frac{t^2}{2}} dt \tag{4.57}$$

For the case of $\mu_0 = 0$, $\mu_1 = 2V_T$, such that

$$P_e = \frac{1}{2}Q\left(\frac{2V_T - V_T}{\sigma_n}\right) + \frac{1}{2}Q\left(\frac{V_T}{\sigma_n}\right) \tag{4.58}$$

Equation 4.58 can be simplified to

$$P_e = Q\left(\frac{V_T}{\sigma_n}\right) \tag{4.59}$$

Substitute $V_T = \mu_1/2$:

$$P_e = Q\left(\frac{\mu_1}{2\sigma_n}\right) \tag{4.60}$$

From Equations 4.1 and 4.60,

$$P_e = Q\left(\frac{P_R R_D}{2\sigma_n}\right) \tag{4.61}$$

Here R is the responsivity and is given by [17]

$$R = \frac{\eta q_c}{h v_l} \tag{4.62}$$

where q_c is the electron charge, h is Planck's constant, η is the quantum efficiency, and v_l is the light frequency.

It is possible to simplify Equation 4.62 such that

$$R = \frac{\eta \lambda}{1.24} \tag{4.63}$$

where λ is measured in units of micrometers.

In practical systems, several noise sources such as thermal, dark current, signal shot noise, and background noise limit the performance of the communication system [7, 17–19]. We describe here the mathematical models of the different noise sources.

The background optical power received by the telescope is obtained by multiplying the background radiance, the telescope aperture size, the telescope FOV, the optical filter bandwidth, and the filter transmission, and it is given by

$$P_{BG} = N_R \cdot \Delta\lambda \cdot T_{\text{filter}}\Omega \cdot \frac{D_R^2}{4} \cdot \pi \qquad (4.64)$$

where D_R is the receiver diameter, $\Delta\lambda$ is the filter bandwidth, T_{filter} is filter transmission, Ω is the receiver FOV, and N_R is the radiance given in units of $\frac{W}{m^2 sr \mu m}$.

The background shot noise is given by

$$\sigma_{BG}^2 = 2q \cdot R_D \cdot P_{BG} B_W \qquad (4.65)$$

where B_W is the receiver electronic bandwidth. The thermal or Johnson noise is given by [7]

$$\sigma_{TH}^2 = \frac{4k_B \cdot T_N}{R_L} B_W \qquad (4.66)$$

where R_L is load resistance, k_B is Boltzmann's constant, and T_N is the noise temperature of the electronic system. The dark current noise is given by

$$\sigma_{DC}^2 = 2 q I_D B_W, \qquad (4.67)$$

where I_D is the photodiode dark current. The signal shot noise is given by

$$\sigma_{SN}^2 = 2q \cdot R_D \cdot P_R B_W, \qquad (4.68)$$

where P_R is the received signal power.

So far in our analysis we have simplified the derivation by assuming that the received power for "0" is equal to zero and that the noise for "1" and "zero" is equal. Without these assumptions we get, for the signal with "1,"

$$\mu_1 = R_D P_{R1} \qquad (4.69)$$

and for the signal with "0,"

$$\mu_0 = R_D P_{R0} \qquad (4.70)$$

In addition, the signal shot noise for "1" is given by

$$\sigma_{SN1}^2 = 2q \cdot R_D \cdot P_{R1} B_W, \qquad (4.71a)$$

and for "0" is given by

$$\sigma_{SN0}^2 = 2q \cdot R_D \cdot P_{R0} B_W \qquad (4.71b)$$

As a result, the total noise variances for "1" and "0" are given by

$$\sigma_1^2 = \sigma_{SS1}^2 + \sigma_{BG}^2 + \sigma_{TH}^2 + \sigma_{DC}^2 \qquad (4.72)$$

and

$$\sigma_0^2 = \sigma_{SS0}^2 + \sigma_{BG}^2 + \sigma_{TH}^2 + \sigma_{DC}^2 \qquad (4.73)$$

respectively.

In the case that the zero and the one noise variance are unequal, the calculation of the bit error rate is complicated. However, an approximation for high signal-to-noise ratio (SNR) was described by [15]

$$\text{BER} \approx Q\left(\frac{\mu_1 - \mu_0}{\sigma_1 + \sigma_2}\right) \qquad (4.74)$$

4.4.2 Pulse Amplitude Modulation

To generalize the OOK modulation method, the pulse amplitude modulation (PAM) was developed. In this method a group of several bits are bound together to form a symbol. All the symbol combinations are defined in the modulation dictionary. The size of the dictionary is

$$S_{\text{dic}} = 2^{S_{\text{sym}}} \qquad (4.75)$$

where S_{sym} is the number of bits in one symbol. Each symbol is mapped to a unique voltage amplitude. Later, this amplitude is converted to an equivalent power that is transmitted by the laser (as presented in Figure 4.12).

The calculation of the error probability is complicated due to the fact that the dictionary size increases from 2 for the OOK case to $S_{\text{dic}} > 2$. It is possible to define an approximation to the symbol error rate if we assume that all noise for all symbols in the dictionary is

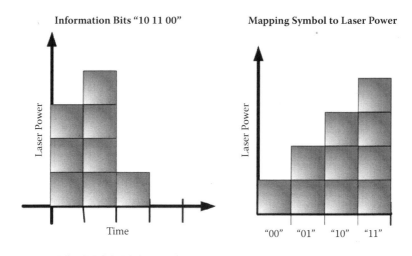

Figure 4.12   Pulse amplitude modulation. (From D. Kedar and S. Arnon, "Non-line-of-sight optical wireless sensor network operating in multi scattering channel," *Appl. Opt.*, vol. 45, pp. 8454–8461, 2006. With Permission.)

identical and defined by σ_N^2 and that the average received power is given by $\bar{P}_R$. In this case we can write the symbol error rate as [19]

$$P_S = \frac{2(M-1)}{M}\left[Q\left(\sqrt{\frac{(\bar{P}_R R)^2 6\log_2 M}{(M^2-1)\sigma_N^2}}\right)\right] \tag{4.76}$$

4.4.3 Pulse Position Modulation

The modulation method named pulse position modulation (PPM) is different in comparison to the previously mentioned methods, OOK and PAM. In OOK and PAM, the information is encapsulated in the amplitude domain, but in PPM the information is encapsulated in the time domain (see Figure 4.13).

The modulation is based on the following guidelines. M bits of information are organized in one symbol. The value of M is a function of the modulation dictionary size. Each symbol is transmitted in T seconds. The period of symbol T is divided into time slots, T_S, where $T = M \cdot T_S$. The symbols are mapped to different time slots according to their value. For example, for dictionary size of $M = 16$ we have 16 time slots, so the symbol 0000 is mapped to the first time slot and 1111 is mapped to the last time slot.

The PPM method requires a larger bandwidth in comparison to PAM and better synchronization (of the order of less than a slot period). On the other hand, it is more robust against fading. The bit error rate for PPM is given by the following equation [14, 20]:

$$\text{BER} = \frac{M}{2(M-1)}$$

$$\times \left[1 - \left[\int_{-\infty}^{\infty} \frac{1}{\sqrt{2\pi}\sigma_1} e^{-\frac{(x-\mu_1)^2}{2\sigma_1^2}} \left(\int_{-\infty}^{x} \frac{1}{\sqrt{2\pi}\sigma_0} e^{-\frac{(y-\mu_0)^2}{2\sigma_0^2}} \, dy \right)^{M-1} dx \right] \right]$$

$$(4.77)$$

where μ_1 and σ_1 are the signal and the noise, respectively, for a slot in which light pulse is received, whereas μ_0 and σ_0 are the signal and the noise, respectively, for a slot without any signal.

4.4.4 The Effect of Turbulence on OOK System Analysis

In this section we analyze the effect of turbulence on a system using OOK modulation methods. From Equations 4.1 and 4.28, the optical signal received at the receiver is obtained by multiplying the transmitter power, telescope gain, and losses and is given by

$$P_R(I) = P_T \eta_T \eta_R \left(\frac{\lambda}{4 \pi Z} \right)^2 G_T \, G_R L_A \cdot I, \qquad (4.78)$$

We consider an intensity modulation/direct detection (IM/DD) receiver using OOK. Therefore, the optical power is converted to electronic signals by a photodetector with a conversion ratio, which is described by the detector responsivity, R (Equation 4.63). We assume that any bias current, dark current, and background light could be removed from the signal without any loss of generality. Integration of the received signal is done for a period of 1 bit. At the end of the integration period, the decision device makes a decision whether the received signal is on or off. The electrical signal without any bias signal before a decision is made is given by y. The noise is assumed to be modeled by additive white Gaussian noise that is statistically

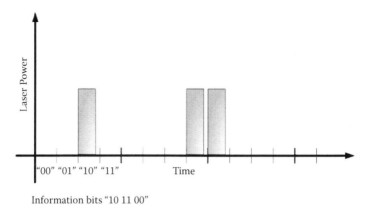

Figure 4.13   Pulse position modulation.

independent of the received signal. The noise has zero mean and covariance σ_N^2. The receiver has knowledge of the turbulence scintillation distribution, as well as of the channel's instantaneous signal fading. It is possible to measure the channel's instantaneous signal fading due to the low frequency of the turbulence signal fading relative to the data rate. The signal y is described by the following conditional probability densities when the transmitted bit is on:

$$P(y|\text{on}, I) = \frac{1}{\sqrt{2\pi\sigma_N^2}} \exp\left(-\frac{\left(y - (RP_R(I))^2\right)}{2\sigma_N^2} \right), \qquad (4.79)$$

or off:

$$P(y|\text{off}) = \frac{1}{\sqrt{2\pi\sigma_N^2}} \exp\left(-\frac{y^2}{2\sigma_N^2} \right), \qquad (4.80)$$

From Equation 4.43, the maximum a posteriori probability (MAP) algorithm decodes the bit $\hat{s}$ as

$$\hat{s} = \max_s \left\{ \frac{P(y|s)P(s)}{P(y)} \right\}, \qquad (4.81)$$

where $P(y|s)$ is the conditional probability that if a bit s is transmitted, a signal amplitude y will be received; $P(s)$ is the probability that an on or off bit is transmitted; and $P(y)$ is the a priori probability of y.

From Equations 4.79–4.81, the likelihood function is given by

$$\Lambda(y, I) = \frac{P(y|\text{on}, I)}{P(y|\text{off})} = \exp\left(-\frac{-2yRP_R(I) + (RP_R(I))^2}{2\sigma_N^2}\right). \quad (4.82)$$

Taking the natural logarithm from both sides of this equation, canceling common factors, and rearranging it yields the following expression:

$$L\Lambda(y, \theta, I) = y - RP_R(\theta, I)/2 \quad (4.83)$$

Then the bit error probability (BEP) is given by

$$\text{BEP} = \int_{-\infty}^{\infty} (P(\text{on}) \cdot P(\text{off/on}, I) + P(\text{off}) \cdot P(\text{on/off}, I)) \cdot f_X(X) \cdot dX, \quad (4.84)$$

where $P(\text{off/on}, I)$ and $P(\text{on/off}, I)$ define the BEP when the on and off bits are transmitted, and are given by

$$P(\text{off/on}, I) = \int_{L\Lambda(y,I)<0} P(y|\text{on}, I) dy \quad (4.85)$$

and

$$P(\text{on/off}, I) = \int_{L\Lambda(y,I)>0} P(y|\text{off}) dy, \quad (4.86)$$

respectively. To simplify Equation 4.84, we define that

$$C\alpha = P_T G_T \frac{R}{2\sqrt{2}\sigma_N} \eta_T \eta_R \left(\frac{\lambda}{4\pi Z}\right)^2 G_R L_A, \quad (4.87)$$

by substituting $(X - E[X])/(\sqrt{2}\sigma_X)$ by v and using the complementary error function $erfc(x)$. Then, the simplified BEP is given by

$$\text{BEP}(\sigma_X,) = \frac{1}{2\sqrt{\pi}} \int_{-\infty}^{\infty} erfc\left(C\alpha \cdot \exp\left(\sqrt{2}\sigma_X v\right)\right) \cdot \exp(-v^2) \cdot dv \quad (4.88)$$

4.4.5 Mitigating Atmospheric Turbulence Effects

From the previous subsection, it is clear that turbulence dramatically affects the performance of any optical wireless system. However, several methods can be used to mitigate the effect of the turbulence. The obvious one is to adapt the transmitter power so that the SNR will be kept at a constant value. However, there is some probability that the beam will wander outside the receiver FOV and, as a result, the power adaptation might not help. In addition, a high-power laser transmitter could create a hazard to the human visual system. Therefore, this solution is limited by the maximum laser exposure regulations. Another method is to use multiple transmitters, which transmit the information to the receiver through multiple uncorrelated spatial channels. As a result, the probability that the turbulence will cause an outage of the communication link reduces dramatically. This method is similar to the concept of diversity in RF communication. The same thing could be done at the receiver side, where the signal from multiple receivers is combined in some optimal way. Another method to mitigate the turbulence effect is to use a bigger aperture in the receiver. A larger aperture in the receiver collects a bigger chunk of the received spot. Because the turbulence is a stochastic process, a bigger chunk of the received spot increases the efficiency of the averaging process.

The last method that we discuss here is adaptive optics. Adaptive optics is a method that senses the wavefront of the received signal and then, based on this information, uses adaptable mirror surfaces to cancel the effect of the turbulence on the wavefront. Mathematically, it is described as a spatial inverse filter for the turbulence.

4.4.6 Performance of an OWC as a Function of Wavelength

OWC is an emerging technology that has many applications in areas such as businesses and offices, short urban wireless links, and on university campuses. However, the major weakness of OWC in terrestrial applications is the threat of downtime caused by bad weather conditions, such as fog and haze. The main mechanism that affects light propagation through fog and haze is scattering by atmospheric aerosols, which deflects randomly the propagating photons in directions other than intended. Consequently, less power is received and the communication system performance is degraded. These phenomena have encouraged many researchers to investigate performance at long wavelengths in the far IR, which minimizes the scattering effect. The main assumption is that in the presence of low attenuation the system availability is much higher. Following this assumption, simulations were carried out for various conditions [21]. The results of these simulations indicate that for total atmospheric attenuation calculations, the preferable transmission wavelength for all visibility conditions was 10 μm, corresponding to 99.8% system availability. This comprises an improvement of 0.2% compared with values for 0.785 μm and 1.55 μm wavelengths. These results were obtained for two U.S. cities at a communication distance of 1 km. An improvement of 0.2% for link availability achieved by the longer wavelengths should be considered with respect to the complexity and cost of quantum cascade laser transmitters and far-IR receivers that require liquid-nitrogen-cooled mercury cadmium telluride detectors. Advanced normalized transmission calculations showed that in clear, hazy, and thin to light fog weather conditions, the preferable transmission wavelengths were within the 9–13 μm long-wave IR band.

However, for thick to dense fog weather conditions, the 11 μm wavelength yields better results. For all the cases studied, the preferred transmission wavelengths were within the 9–13 μm band, with superior transmissivity near the 11 μm wavelength. Therefore, the use of quantum cascade laser transmitters as a unique long-wave IR laser source is recommended. Performance analysis of communication systems, as determined by digital signal-to-noise ratio (DSNR), was carried out taking into account the dominant noise sources. For a background-noise-limited system, the best performance was obtained within the 0.4–0.7 μm visible band, owing to its much lower background radiation level than that of the 7–15 μm wavelength band. For a thermal-noise-limited system in clear, hazy, and thin to light foggy weather conditions, however, similar system performance was observed for the entire spectrum, whereas in thick to dense foggy weather conditions, better system performance was obtained within the 8–13 μm long-wave IR band. The advantage of longer IR wavelengths relative to shorter wavelengths in dense fog weather conditions for a 200 m terrestrial link path was evaluated in a recent experiment in which clearly better performance of an 8.1 μm quantum cascade laser transmitter compared with a 0.85 μm near-IR laser was reported. All these results underscore the problematic nature of free-space transmission through dense fog. The influence of atmospheric turbulence on link performance lessens at longer wavelengths. Moreover, if the system is designed with multiple transmitters and wide receiver aperture, scintillation fading can be reduced. In the studies reported in this section, all the calculations were performed for a 1 km terrestrial link path. Obviously, for a longer distance OWC, link system performance worsens. In the future, research should examine field experiment results, advanced adaptive methods, or hybrid OWC-RF systems that operate properly under such conditions. In all cases, the choice of OWC operation wavelength should take eye-safety requirements into consideration.

As light propagates through the atmosphere, it will be absorbed or scattered by molecules and aerosols [5, 7, 22] (see also Chapter 2, where these effects are fully described). Because these mechanisms are highly wavelength sensitive, the transmission of light at two different wavelengths through a given propagation channel differs significantly. Usually, radiation wavelengths that are heavily absorbed by atmospheric particles are avoided in optical wireless communication systems to minimize beam attenuation and consequent power requirements. However, background radiation at the transmitted wavelength, particularly due to solar radiance during daytime operation, produces background shot noise that contaminates the desired SNR. To combat background shot noise, the receiver FOV and the filter are designed to be narrow [22]. However, we can use the fact that the spectrum of solar irradiance reaching the ground is far from uniform. Notably, almost all the solar radiation in the spectral region around 200–280 nm is absorbed by ozone in the upper atmosphere. Hence, virtually no background noise would be encountered when the transmission wavelength is in this region, which is known as "solar-blind ultraviolet," and very large FOV receivers can be used. Additionally, considerable scatter by atmospheric particles occurs at very short wavelengths. This enables the establishment of a non-line-of-sight (NLOS) communication regime, in which the atmospheric particles act as reflective elements and close a link between nonaligned transmitter-receiver pairs. (See Figure 4.14.)

It can be seen clearly from Figure 4.14 that, for a given transmitter beam divergence and transmitter-receiver separation, a larger receiver

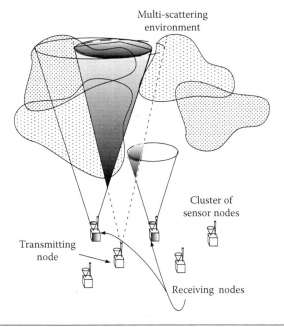

Figure 4.14   Possible scenario; microsensors strewn ad hoc on the ground self-orientate to face vertically upward. Communication is achieved by virtue of backscatter from multi-scattering environment. (From D. Kedar and S. Arnon, "Non-line-of-sight optical wireless sensor network operating in multi scattering channel," *Appl. Opt.*, vol. 45, pp. 8454–8461, 2006. With Permission.)

FOV would result in a larger intersection of the beam and FOV cones, and more backscattered optical power from the transmitter would reach the receivers. Similarly, an increased transmitter beam divergence would enlarge the intersection area. In conclusion, an optical wireless system operating in the solar-blind ultraviolet spectral range could achieve NLOS communication by means of backscatter from atmospheric particles and use large FOV receivers to collect the scatter power.

References

1. D. Kedar and S. Arnon, "Urban optical wireless communication network: The main challenges and possible solutions," *IEEE Commun. Magazine*, vol. 42, pp. S2–S7, 2004.
2. S. Arnon, "Optical wireless communication," in *Encyclopedia of optical engineering (EOE)*, R. G. Driggers, Ed., Marcel Dekker, New York, pp. 1866–1886, 2003.

3. B. J. Klein and J. J. Degnan, "Optical antenna gain. 1: Transmitting telescopes," *Appl. Opt.*, vol. 13, pp. 2134–2141, 1974.

4. J. J. Degnan and B. J. Klein, "Optical antenna gain. 2: Receiving telescopes," *Appl. Opt.*, vol. 13, no. 10, pp. 2397–2401, 1974.

5. E. J. McCartney, *Optics of the Atmosphere*, John Wiley & Sons, New York, 1976.

6. I. I. Kim, B. McArthur, and E. Korevaar, "Comparison of laser beam propagation at 785 nm and 1550 nm in fog and haze for optical wireless communications," E. J. Korevaar, Ed., *Proc. SPIE Opt. Wireless Commun. III*, vol. 4214, pp. 26–37, 2000.

7. N. S. Kopeika, *A System Engineering Approach to Imaging*, SPIE Optical Engineering Press, Bellingham, WA, 1998.

8. W. L. Wolf and G. Zissis, Eds., *The Infrared Handbook*, Office of Naval Research, Department of the Navy, Arlington, VA, 1985.

9. S. Karp, R. M. Gagliardi, S. E. Moran, and L. B. Stotts, *Optical Channels Fibers, Clouds, Water, and the Atmosphere*, Plenum, New York, 1988.

10. L. C. Andrews and R. L. Philips, *Laser Beam Propagation through Random Media*, SPIE Optical Engineering Press, Bellingham, WA, 1998.

11. X. Zhu and J. M. Kahn, "Free-space optical communication through atmospheric turbulence channels," *IEEE Trans. Commun.*, vol. 50, pp. 1293–1300, 2002.

12. M. Uysal, L. Jing, and Y. Meng, "Error rate performance analysis of coded free-space optical links over gamma-gamma atmospheric turbulence channels," *IEEE Trans. Wireless Commun.* vol. 5, pp. 1229–1233, 2006.

13. M. A. Al-Habash, L. C. Andrews, and R. L. Phillips, "Mathematical model for the irradiance probability density function of a laser beam propagating through turbulent media," *Opt. Eng.*, vol. 40, pp. 1554–1562, 2001.

14. R. M. Gagliardi and S. Karp, *Optical Communication*, John Wiley & Sons, New York, 1995.

15. S. B. Alexander, *Optical Communication Receiver Design*, SPIE Optical Engineering Press, Bellingham, WA, 1997.

16. G. P. Agrawal, *Fiber Optic Communication*, 2nd ed., John Wiley & Sons, New York, 1997.

17. W. B. Jones, *Introduction to Fiber Communication Systems*, Holt, Rinehart & Winston, Oxford University Press, New York, 1988.

18. G. Keiser, *Optical Fiber Communication*, 2nd ed., McGraw-Hill, New York, 1991.

19 J. G. Proakis, *Digital Communication*, 4th ed., McGraw-Hill, New York, 2001.

20. C. C. Chen and C. S. Gardner, "Impact of random pointing and tracking errors on the design of coherent and incoherent optical intersatellite communication links," *IEEE Trans. Commun.*, vol. 37, pp. 252–260, 1989.

21. H. Manor and S. Arnon, "Performance of an optical wireless communication system as a function of wavelength," *Appl. Opt.*, vol. 42, pp. 4285–4294, 2003.

22. D. Kedar and S. Arnon, "Non-line-of-sight optical wireless sensor network operating in multi scattering channel," *Appl. Opt.*, vol. 45, pp. 8454–8461, 2006.

23. D. Kedar and S. Arnon, "Backscattering-induced crosstalk in WDM optical wireless communication," *IEEE/OSA J. Lightwave Tech.*, vol. 23, pp. 2023–2030, 2005.

24. M. Aharonovich and S. Arnon, "Performance improvement of optical wireless communication through fog by a decision feedback equalizer," *J. Optic. Soc. Am. A*, vol. 22, pp. 1646–1654, 2005.

5

CHANNEL AND SIGNAL DATA PARAMETERS IN ATMOSPHERIC OPTICAL COMMUNICATION LINKS

In recent decades, there has been increasing demand for high-data-rate transmission over wireless optical communication links [1–5]. This interest stems from the advantages of using optical systems over conventional radiofrequency (RF) systems, such as less mass, power, and volume along with high gain and narrow band. Unfortunately, the performance of optical systems can be corrupted significantly by atmospheric turbulence caused by strong multipath fading due to multiple scattering of optical waves on various turbulent structures.

Atmospheric media vary randomly in time and space, so that the amplitude and phase of wave propagation may similarly fluctuate randomly in these domains [6–10]. Recently, considerable effort has been devoted to studying influences of atmospheric effects on waves, optical or radio, propagating along or through the atmosphere [11–22]. Many other cases of degrading propagation phenomena occur within atmospheric wireless communication links, such as absorption by gases, molecular absorption, scattering and attenuation by hydrometeors (fog, smoke, rain, snow, and so on), attenuation by atmospheric turbulence, and scattering by irregularities of refractive index [23–34] (see also Chapters 1–3). The atmosphere of the Earth is a rather opaque, quite inhomogeneous, and increasingly refractive medium through which most radio and optical signal experiences are propagated [13, 16]. As for noises in atmospheric communication links and effects on the data stream parameters passing through such channels, the main source of them is turbulence, which causes signal scintillation or *fading* [34–36]. Atmospheric turbulence is the cause of random fluctuations of signal intensity and phase [21–30].

Thus, as was mentioned in Chapters 1–3, the turbulent atmosphere causes the intensity of a wave beam to fluctuate or scintillate and causes beam wander, distortion, and random displacement. The optical effects of atmospheric turbulence depend primarily on the refractive index structure parameter C_n^2. Among these atmospheric phenomena, atmospheric turbulence plays a significant role in the creation of the distortion to the received signal [16–20].

To achieve high performance along with high data rate and an increase in channel capacity parameters including minimization of bit error rate (BER), it is important to decrease the impact of atmospheric processes, which influence signal propagation within the communication link. The purpose of this chapter is to develop a better evaluation of the expected communication parameters in the atmospheric communication channel by measuring refraction index parameter C_n^2 via angle-of-arrival (AOA) fluctuations of the received optical signal, which are related to the intensity of the atmospheric turbulence (from weak to strong). The relation between gamma-gamma probability density function (PDF), usually used in atmospheric optical communication links, and Ricean PDF, usually used in radio multipath fading communication links, using the relation between the signal scintillation parameter σ_I^2 and the Ricean K parameter, is found to unify the stochastic approach for both kinds of atmospheric communication links, optical and radio. Below, we present the corresponding formulas via Ricean parameter K, which were evaluated for calculation of the data stream parameters based on estimations of the refraction index parameter C_n^2 using experimental data from the corresponding measurements.

5.1 Irradiance PDF

Over the years, many irradiance PDF models have been proposed to describe atmospheric turbulence fluctuations for wireless communication channels. The difference between the models is in the turbulence intensity (weak to strong) [7, 16, 17].

The lognormal PDF is usually related to weak turbulence that gives close evaluations of the expected behavior of irradiance fluctuations and of PDF tails, which have important consequences on radar and optical communication channel parameters [7, 9, 16, 17, 22]. When atmospheric turbulence is strong and includes multiple self-interference effects, the lognormal statistics are no longer accurate, and the zero-mean Gaussian with negative exponential distribution, which defines the limit distribution for irradiance fluctuations [31], can yield a better evaluation.

5.1.1 Gamma-Gamma Distribution

The more common and well-known model to characterize the irradiance fluctuations caused by atmospheric turbulence that takes into account also the assumed modulation processes of wave propagation is the *gamma-gamma* distribution model [7, 29, 32].

The gamma-gamma distribution can be derived from the modulation process, where the double stochastic negative exponential distribution is directly related to the atmospheric parameters (the large scale, L_0, and the small scale, l_0), by describing the large and the small scales as gamma distributions [16, 32]:

$$p_x(x) = \frac{\alpha(\alpha x)^{\alpha-1}}{\Gamma(\alpha)} \exp(-\alpha x), \quad \text{for} \quad x > 0, \quad \alpha > 0, \quad (5.1)$$

$$p_y(y) = \frac{\beta(\beta y)^{\beta-1}}{\Gamma(\beta)} \exp(-\beta y), \quad \text{for} \quad y > 0, \quad \beta > 0.$$

where x and y are mean irradiance, $x = y = \langle I \rangle$, which are random quantities, and I is the irradiance. In Equation 5.1, Γ is the gamma function and parameters α and β correspond to weak and strong fluctuations, respectively. First, by fixing x and then writing $y = I/x$, we obtain the conditional PDF:

$$p_y(I \mid x) = \frac{\beta(\beta I/x)^{\beta-1}}{x\Gamma(\beta)} \exp(-\beta I/x), \quad \text{for} \quad I > 0 \quad (5.2)$$

where x is the conditional mean value of the irradiance intensity. To obtain the unconditional irradiance PDF we have to average $p_y(I|x)$ from Equation 5.2 over the gamma distribution (Equation 5.1), which eventually leads to the gamma-gamma distribution:

$$p(I) = \int_0^\infty p_y(I \mid x) p_x(x) dx$$

$$\qquad\qquad\qquad\qquad\qquad\qquad\qquad\qquad (5.3)$$

$$= \frac{2(\alpha\beta)^{(\alpha+\beta)/2}}{\Gamma(\alpha)\Gamma(\beta)} I^{[(\alpha+\beta)/2]-1} K_{\alpha-\beta}[2(\alpha\beta I)^{1/2}], \quad \text{for} \quad I > 0,$$

where $K_{\alpha-\beta}[\bullet]$ represents the Bessel function of the second kind of $(\alpha-\beta)$ order. This K or gamma-gamma distribution can be divided into two regimes, considering parameters (α and β), where α represents the effective number of the large-scale random variable and β is related to that of the small scale. When turbulence is weak, effective numbers of scale sizes are either smaller or much larger relative to the first Fresnel zone [9, 22, 32].

Strong turbulence can be characterized by β decreasing beyond the focusing regime and approaching saturation, where $\beta \to 1$ means that the number of small-scale cells is reduced to one. On the other hand, the number of discrete refractive scatters α increases under strong turbulence conditions [7, 9].

Nevertheless, the gamma-gamma distribution can be approached as a negative exponential distribution [31, 32]. The large- and small-scale parameters of mean turbulence may be written:

$$\alpha = \frac{1}{\sigma_x^2}, \quad \beta = \frac{1}{\sigma_y^2} \tag{5.4}$$

where σ_x^2 and σ_y^2 are the normalized variances of x and y, respectively. Eventually, the total scintillation index is [16]

$$\sigma_I^2 = \frac{1}{\alpha} + \frac{1}{\beta} + \frac{1}{\alpha\beta} \tag{5.5}$$

This distribution will be used for numerical simulation of the problem and discussions on the results of computations presented.

5.1.2 Ricean Distribution

To estimate the contribution of each component, the dominant [or line-of-sight (LOS)] and the multipath [or non-line-of-sight (NLOS)], to the resulting signal at the receiver, the Ricean parameter K is usually introduced as a ratio between these components [6, 34]. The Ricean PDF distribution of the signal strength or voltage envelope r can be defined as [6, 33, 34]

$$\text{PDF}(r) = \frac{r}{\sigma^2} \exp\left\{-\frac{r^2 + A^2}{2\sigma^2}\right\} \cdot I_0\left(\frac{Ar}{\sigma^2}\right), \quad A > 0, r \geq 0 \tag{5.6}$$

where A denotes the peak strength or voltage of the dominant component envelope, σ is the standard deviation of the signal envelope, and $I_0(\cdot)$ is the modified Bessel function of the first kind and zero order. Following definitions introduced in References 6 and 34, we can present the parameter K as the ratio between the dominant and the multipath component powers. Thus, it is given by References 6, 33, and 34 as the ratio of coherent and incoherent (multipath) components of the total received signal:

$$K = \frac{A^2}{2\sigma^2} = \frac{I_{co}}{I_{inc}} \tag{5.7}$$

Using Equation 5.7, we can rewrite Equation 5.6 as a function only of K, as was done in References 6 and 34:

$$\text{PDF}(x) = \frac{r}{\sigma^2} \exp\left\{-\frac{r^2}{2\sigma^2}\right\} \cdot \exp(-K) \cdot I_0\left(\frac{r}{\sigma}\sqrt{2K}\right) \tag{5.8a}$$

from which for $K = 0$, $\exp(-K) = 1$ and $I_0(0) = 1$, the worst case of the fading channel, described by Rayleigh PDF, follows, when there is no LOS signal and PDF equals

$$\text{PDF}(x) = \frac{r}{\sigma^2} \exp\left\{-\frac{r^2}{2\sigma^2}\right\} \tag{5.8b}$$

Conversely, in a situation of good clearance between two terminals with no multipath components (i.e., when $K \rightarrow \infty$), the Ricean fading approaches a Gaussian one obtaining a "Dirac-delta-shaped" PDF [6, 33, 34].

Usually, wireless communication systems use well-known modulation techniques for encoding the received signal [33]. While applying the coherent detection process for encoding of information into the amplitude, frequency, or phase of the transmitted signal (as a carrier of this information), commonly known modulation techniques often used are amplitude shift keying (ASK), frequency shift keying (FSK), and phase shift keying (PSK), when the modulation follows changes of amplitude, frequency, and phase, respectively [33, 34]. We will describe the case of ASK modulation, which deals with the signal strength or power as a function of time. Because the optical signal can be considered as a carrier of digital information, as a set of bits, effects of multipath fading in the optical communication channel lead to errors in bits characterized by the special parameter defined as BER. Thus, using the Ricean distribution, we can determine the probability of bit error occurring in the multipath channel operating with ASK modulation by the following formula [7, 32, 34]:

$$P_r(e) = \frac{1}{\sigma^2} \int_{r_T}^{\infty} r e^{-\left[\frac{r^2}{2\sigma_N^2}\right]} dr = e^{-\left[\frac{r_T^2}{2\sigma_N^2}\right]} \qquad (5.9)$$

where $P_r(e)$ represents the evaluated probability of bit error, σ_n^2 is the intensity of interference at the optical receiver (usually determined as the multiplicative noise [7, 32, 34]), and r_T determines the threshold between detection without multiplicative noise (defined as a "good case" [34]) and with multiplicative noise (defined as a "bad case" [34]).

5.2 Key Parameters of Data Stream in Optical Channels with Fading

5.2.1 BER of Optical Channel

Expected probability of error in optical ASK modulation systems was suggested by Andrews and Phillips [7, 16] as a function of PDF of random variable s, $p_i(s)$, which represents the well-known gamma-gamma distribution [7, 16]:

$$P_r(e) = \frac{1}{2} \int_0^{\infty} p_i(s) \cdot erfc\left(\frac{\langle SNR \rangle \cdot s}{2\sqrt{2}\langle i_s \rangle} \right) ds \qquad (5.10)$$

Here, as above, $P_r(e)$ is the evaluated probability of error, $erfc(*)$ is the error function probability [32], and $\langle i_s \rangle$ is the average intensity of the received signal.

In our investigations we use the well-known relations between σ_I^2, σ_x^2, and σ_y^2 [7, 16] with C_n^2 and β_0^2 in order to find the relation between the signal parameter σ_I^2 and the channel parameters C_n^2 [7, 16, 35].

$$\sigma_x^2 = \exp\left\{ \frac{0.49\beta_0^2}{\left(1 + 0.18d^2 + 0.56\beta_0^{12/5}\right)^{7/6}} \right\} - 1 \qquad (5.11a)$$

$$\sigma_y^2 = \exp\left\{\frac{0.51\beta_0^2}{(1+0.9d^2+0.62\beta_0^{12/5})^{5/6}}\right\} - 1 \qquad (5.11b)$$

where β_0^2 is the corresponding level of turbulence, defined as

$$\beta_0 = 0.5C_n^2 k^{7/6} L^{11/6} \qquad (5.12)$$

and d represents the total diameter of an array of lenses at the receiver:

$$d = (kD^2/4L)^{1/2} \qquad (5.13)$$

where k is the optical wavenumber parameter, which equals $2\pi/\lambda$, and λ is wavelength; β_0^2 corresponds to the level of turbulence, thus $\beta_0^2 = 0.1$ for weak fluctuation and $\beta_0^2 = 4$ for strong turbulence; and d is the normalized aperture parameter of the receiver, so that $d = 0$ is for a point receiver and $d = 10$ is for a large collecting lens [16, 35].

Using the previous definitions, according to References 16 and 35, we can rewrite Equation 5.10 in the following form:

$$P_r(e) = \frac{1}{2}\int_0^\infty erfc\left(\frac{SNR \cdot s}{2\sqrt{2\langle i_s \rangle}}\right)\frac{2(\alpha\beta)^{(\alpha+\beta)/2}}{\Gamma(\alpha)\Gamma(\beta)\langle i_s \rangle}\left(\frac{s}{\langle i_s \rangle}\right)^{\frac{(\alpha+\beta)}{2}-1}$$

$$\times K_{(\alpha-\beta)}\left(2\sqrt{\frac{\alpha\beta s}{\langle i_s \rangle}}\right) ds \qquad (5.14)$$

where $\langle i_S \rangle$ represents the average intensity of the received signal and s is the random statistically independent variable of integration.

5.2.2 Channel Capacity and Spectral Efficiency

To characterize an optical communication channel and determine its quality, we use several parameters that allow us to estimate the quality of the received data versus the sent data and parameters that describe the quality of the transmitted signal along the transmission channel, such as the capacity, C, and the spectral efficiency, $\tilde{C}$. The capacity is defined as a maximum rate of information data stream within the communication channel, and the spectral efficiency is a ratio of the capacity and the bandwidth of the channel.

To find all these key parameters of the channel and information data stream, we present two approaches: the classical approach, based on the well-known probabilistic theory [16, 32, 33, 37–43], and the proposed approximate approach, based on a multi-parametric stochastic model [34–36, 44].

5.2.2.1 Classical Approach   Following References 32 and 37–42, we look at the following classical model of the wireless channel and present the resulting signal passing such a channel as

$$r(t) = h(t; x)s(t) + n(t) \tag{5.15}$$

where $h(t; x)$ is the channel gain and $n(t)$ is the white (additive) noise. The channel gain, $h(t; x)$, is a random complex variable, which we define as having mean $\mu(x)$, which also changes as a function of space. We shall thus describe the channel as

$$h(x, t) = \mu(x) + h_r(t) \tag{5.16}$$

where $h_r(t)$ is a zero-mean complex normal variable with standard deviation σ_h. The mean, $\mu(x)$, is also a random variable with respect to x and can be considered to have a mean $\bar{\mu}(x)$ and a random part $\mu_r(x)$, which is lognormally distributed with zero mean. Then, the signal-to-noise ratio (SNR) is given by

$$\rho = \frac{|h|^2}{\sigma^2} \tag{5.17}$$

where σ^2 is the noise standard deviation.

The capacity of a communication channel can be defined as the traffic load of data in bits per second. It is acceptable to use the Shannon-Hartley formula to calculate the capacity of the channel with additive white Gaussian noise (AWGN) [32, 37–42]:

$$C(r) = B_w \log_2 [1 + \rho] \tag{5.18a}$$

where ρ is the SNR. Then, the spectral efficiency can be presented as

$$\tilde{C}(r) = \frac{C(r)}{B_w} = \log_2[1 + \rho] \tag{5.18b}$$

We rewrite Equation 5.18a by introducing in it only the density, N_0, of AWGN (in W/Hz) inside the channel with a large bandwidth,

B_w (in Hz), of noise spectrum:

$$C = B_w \log_2\left[1 + \frac{S}{N_0 B_w}\right] \qquad (5.18c)$$

where now the power of additive noise in the AWGN channel is $N_{add} = N_0 B_w$.

In a general case when there are both additive and multiplicative (flat) noises in the optical communication link with fading, the average spectral efficiency can be presented, following References 16 and 43, as

$$\tilde{C} = \int_{|h|}\int_{\mu} \log_2\left(1 + \frac{|h|^2}{\sigma^2}\right) p(h, \mu(x))\, dh\, dx \qquad (5.19)$$

Here the joint probability distribution function of $|h|$ and $\mu(x)$ is given by [16, 43]

$$p(h, \mu(x)) = p[|h| \,||\, \mu(x)]\, p(\mu(x))$$

$$= \frac{1}{\sqrt{2\pi\sigma_\mu^2}}\, e^{-\frac{\log\mu(x) - \log\bar{\mu}}{2\sigma_\mu^2}}\, \frac{|h|}{\sigma_h^2} \qquad (5.20)$$

$$\times e^{-\frac{|h|^2 + \mu(x)^2}{2\sigma_h^2}}\, I_0\left(\frac{|h|\,\mu(x)}{\sigma_h^2}\right)$$

where $I_0(w)$ is the Bessel function of zero order; all other parameters have been introduced.

5.2.2.2 Approximate Approach We now propose an approximate approach, which provides a satisfactory explanation for the relations between the parameters of the signal, such as K factor, the key parameters of the channel, such as capacity and spectral efficiency, and the parameters of the data stream, such as bit rate and BER. This approach is based on introduction of a "multiplicative noise" term in addition to the additive noise term.

Let us prove this approach following Reference 35. Our simple approach can be used in cases of dynamic communication channels with fast fading, flat or multiplicative, where an additional source of noise, called multiplicative noise, can be introduced [6, 8, 9].

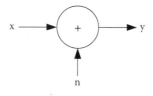

Figure 5.1   Basic communication channel description.

To prove this approach, we use the basic definition of the capacity determined via entropy in [32, 33, 41–43, 45]

$$C = \text{Max}(H(x) - H_y(x)) = \text{Max}(H(y) - H_x(y)) \qquad (5.21)$$

where $H(x)$ is the continuous entropy of the source and $H(y)$ is the continuous entropy of the receiver (see Figure 5.1). Here also $H_y(x)$ is the conditional entropy of the given x at the receiver y (i.e., the average entropy of x for each value of y), and, accordingly, $H_x(y)$ is the conditional entropy of the given y for the source x (i.e., the average entropy of y for each value of x).

The channel is defined in such a way that $y = x + n$ (see Figure 5.1), and the capacity is defined as follows [32, 33, 42, 43]:

$$C = \underset{p(x)}{\text{Max}}\left\{H(y) - H(n)\right\} \qquad (5.22)$$

The entropy of a continuous distribution can be defined as

$$H(x) = -\int_{-\infty}^{\infty} p(x) \log(p(x)) dx.$$

Let us now assume that x is a Gaussian random variable. Its entropy, according to the definition introduced in References 5 and 8, is

$$H(x) = -\int_{-\infty}^{\infty} p(x) \log(p(x)) dx$$

$$= -\int_{-\infty}^{\infty} \frac{1}{\sqrt{2\pi}\sigma} e^{-(x^2/2\sigma^2)} \log\left(\frac{1}{\sqrt{2\pi}\sigma} e^{-(x^2/2\sigma^2)}\right) dx \qquad (5.23)$$

$$= \log\left(\sqrt{2\pi e}\sigma\right)$$

In the case of a Gaussian *independent, identically distributed* (i.i.d.) *channel*, the PDF function of the received signal is the convolution of the PDFs of the input signal and the noise; i.e.,

$$\text{PDF}(y) = x \oplus n = \int_{-\infty}^{\infty} \frac{1}{\sigma_1\sqrt{2\pi}} \cdot e^{-\frac{t^2}{2\sigma_1^2}} * \frac{1}{\sigma_2\sqrt{2\pi}} \cdot e^{-\frac{(\tau-t)^2}{2\sigma_2^2}}\, dt$$

(5.24)

$$= \frac{1}{(\sigma_1 + \sigma_2)\sqrt{2\pi}} \cdot e^{-\frac{t^2}{2(\sigma_1^2 + \sigma_2^2)}}$$

Thus, the entropy of the receiver in the case of the i.i.d. Gaussian channel is

$$H(y) = -\int_{-\infty}^{\infty} \text{PDF}(y)\log(\text{PDF}(y)) = \log\left(\sqrt{2\pi e}(\sigma_1 + \sigma_2)\right)$$

(5.25)

$$= \log\left(\sqrt{2\pi e}(S + N)\right)$$

The capacity of such a channel, according to Equation 5.22, is

$$C = \underset{p(x)}{\text{Max}}\{H(y) - H(n)\} = B_\omega \log\left(\sqrt{2\pi e}(S + N)\right) - B_\omega \log\left(\sqrt{2\pi e}(N)\right)$$

$$= B_\omega \log\left(\frac{2\pi e(S + N)}{2\pi e(N)}\right) = B_\omega \log\left(1 + \frac{S}{N}\right)$$

(5.26)

It is well known that the sum of i.i.d. Gaussian random variables is also a Gaussian random variable, where the mean value μ is the sum of the mean values (i.e., $\mu = \mu_1 + \mu_2$) and the variance σ^2 is the sum of the variances (i.e., $\sigma^2 = \sigma_1^2 + \sigma_2^2$) [33].

Thus, if in the wireless channel there are two i.i.d. Gaussian noises, N_1 and N_2, we can consider them as a new total noise: $N_{\text{tot}} = N_1 + N_2$. In this case the entropy will simply have the additive form:

$$H(N_{\text{tot}}) = H(N_1 + N_2).$$

Thus, the capacity of such a channel is

$$C = B_\omega \log\left(1 + \frac{S}{N_{tot}}\right) = B_\omega \log\left(1 + \frac{S}{N_1 + N_2}\right) \qquad (5.27)$$

Taking into account multiplicative noise influence with effective power density N_{mul} as one of Gaussian-like noises, we can rewrite Equation 5.18c as

$$C = B_\omega \log_2\left[1 + \frac{S}{N_0 B_\omega + N_{mul} B_\Omega}\right] \qquad (5.28)$$

where B_Ω is the frequency bandwidth of the multiplicative noise.

We should mention here that this procedure of replacing Equation 5.18c with Equation 5.28 can be performed only because of the similarity between the Ricean distribution and the Gaussian distribution for large K, and the fact that as the Ricean K parameter grows, the Ricean distribution approaches the Gaussian distribution in the limit of $K \gg 1$.

Using the approximate formula in Equation 5.28 for the capacity (or for the spectral efficiency of the channel, $\tilde{C} = C/B_w$), we can now simplify the expressions obtained for capacity, spectral efficiency, and BER, which are usually used in wireless communications [32, 37–39, 42–45]. Again, Equation 5.28 is valid when the LOS component of the total signal inside a channel exceeds the NLOS component; that is, when the K factor is greater than unity. This condition is fully satisfied for optical atmospheric links.

Now we will investigate each key parameter separately. First, we rewrite the formula in Equation 5.28, describing the capacity of the channel as

$$C = B_w \log_2\left(1 + \frac{S}{N_{add} + N_{mul}}\right) = B_w \log_2\left(1 + \left(\frac{N_{add}}{S} + \frac{N_{mul}}{S}\right)^{-1}\right)$$

$$(5.29)$$

where, as before, $\frac{S}{N_{add}} = \text{SNR}_{add}$ and $\frac{S}{N_{mul}} = \frac{I_{co}}{I_{inc}}$. Here, we also use the previous definition of K as the ratio $K = I_{co}/I_{inc}$. Using these notations, we finally get the capacity as a function of the K factor and the signal-to-additive-noise ratio (SNR_{add}):

$$C = B_w \log_2\left(1 + \left(\text{SNR}_{add}^{-1} + K^{-1}\right)^{-1}\right) = B_w \log_2\left(1 + \frac{K \cdot \text{SNR}_{add}}{K + \text{SNR}_{add}}\right)$$

(5.30)

Consequently, it is easy to obtain from Equation 5.18b the spectral efficiency of the channel:

$$\tilde{C} = \frac{C}{B_w} = \log_2\left(1 + \frac{K \cdot \text{SNR}_{add}}{K + \text{SNR}_{add}}\right)$$

(5.31)

where the bandwidth, B_w, changes according to the system under investigation. As we compare the two approaches, classical and approximate, it will be shown that Formulas 5.30 and 5.31, compared with 5.19, are valid when the K factor is larger than SNR_{add}.

Now, we will express K as a function of C using Equation 5.31. Finally, we get

$$K = \frac{\text{SNR}_{add}\left(2^{\frac{C}{B_w}} - 1\right)}{\text{SNR}_{add} - \left(2^{\frac{C}{B_w}} - 1\right)} = \frac{\text{SNR}_{add}\left(2^{\tilde{C}} - 1\right)}{\text{SNR}_{add} - \left(2^{\tilde{C}} - 1\right)}$$

(5.32)

This equation is an important result, because it gives a relation between the spectral efficiency of the multipath communication channel caused by fading phenomena and the BER inside such a channel.

Finally, we can relate the strength of the scintillation, which is characterized by the normalized intensity variance, $\langle\sigma_I^2\rangle$, called scintillation index (see also Chapter 2), with the measured intensity fluctuations, I, as [9, 16, 19]

$$\langle\sigma_I^2\rangle = \frac{\langle[I - \langle I\rangle]^2\rangle}{\langle I\rangle^2} = \frac{I_{inc}^2}{I_{co}^2} \equiv K^{-2}$$

(5.33)

where I_{co} and I_{inc} are the coherent and incoherent components of the total signal intensity.

5.3 Modeling of Key Parameters of the Channel and Information Data

In the previous section we described several models and parameters to characterize atmospheric turbulence properties, which affect optical communication link quality. In this section we show graphical relationships between those communication parameters, K parameter and BER; the channel spectral efficiency, $\tilde{C}$; and levels of atmospheric turbulence, β_0.

First, we compare two approaches: the classical and the approximate. For these purposes we analyze the changes of the capacity as a function of K (i.e., as a function of different conditions of the channel), using Equations 5.30 and 5.31, and examining the spectral efficiency for three different typical values of SNR_{add}, which is shown in Figure 5.2. The dashed curves represent results of computations according to the classical approach described by Equation 5.19. Continuous curves describe results of computations based on the approximate approach using Equation 5.31. All curves are depicted for different values of signal-to-additive-noise ratio, denoted as before by SNR_{add}. The presented illustrations make it clear that with the increase in SNR_{add} (from 1 to 10 dB), the spectral efficiency is increased by more than three times. It is also clearly seen that both approaches are close when $K \gg 1$. This effect is more vivid for the additive noise, which is less than parameter K in decibels; i.e., for $SNR_{add} \leq 10$ dB.

Now, we can enter into detailed analysis of the key parameters of the optical atmospheric channel and the information data, based on the theoretical approach already proposed.

Graphical representation of evaluation results is based on a developed algorithm consisting mainly of Equation 5.14 based on

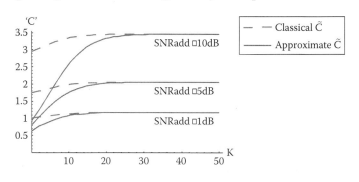

Figure 5.2   Spectral efficiency as a function of K factor.

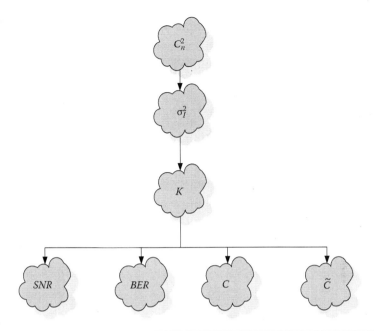

Figure 5.3 The model flow diagram. (From A. Tiker, N. Yarkoni, N. Blaunstein, A. Zilberman, and N. Kopeika, "Prediction of data stream parameters in atmospheric turbulent wireless communication links," *Appl. Opt.*, vol. 46, no. 2, pp. 190–199, 2007. With Permission.)

5.10–5.13 and 5.30–5.33. The flow diagram shown in Figure 5.3 easily can describe our algorithm. First, experimentally we measure the refractive index structure parameter, C_n^2, which characterizes the strength of the atmospheric turbulence. In addition, parameter K, which represents the ratio between LOS and NLOS components of the communication link, is evaluated. The calculation of the energy of the LOS component is the measured transmitted energy divided by the attenuation along the propagation path. At the same time, the calculation of the NLOS energy is taken from the measured beam scintillation/fluctuation energy, σ_I^2, according to Equation 5.33. The latter is evaluated based on the relation between σ_I^2, α, β, and β_0^2 as a function of C_n^2 (see Equations 5.10–5.13). Considering these parameters, finally, the SNR is calculated. Then, using Equations 5.14 and 5.30–5.31, the BER, the capacity, and the channel spectral efficiency can be evaluated, respectively.

Therefore, based on the corresponding flow diagram in Figure 5.3, we obtained the following results. First, we evaluated β_0^2 based on the modeled parameter C_n^2 according to References 24, 36, and 43.

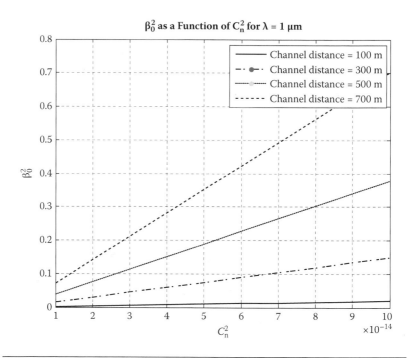

Figure 5.4 β_0^2 as a function of C_n^2 for different ranges of the channel.

(See also results discussed in Chapters 1 and 2.) Figure 5.4 shows the dependence of the turbulence parameter β_0^2 versus the parameter C_n^2, based on Equation 5.12, for different ranges, L, of the channel.

As can be seen in Figure 5.4, the bottom curve is for the range of 100 m and the top curve is for the range of 700 m. Results shown in Figure 5.4 correspond to different levels of atmospheric turbulence: from weakest (at small ranges) to strongest (at large ranges). The nominal values of the refractive index structure parameter used in our computation are from $C_n^2 \approx 10^{-15}$ m$^{-2/3}$ to $C_n^2 \approx 10^{-13}$ m$^{-2/3}$, according to References 24 and 36, with a mean value of $\langle C_n^2 \rangle \approx 10^{-15}$ m$^{-2/3}$. As follows from the illustration in Figure 5.4, the rate of β_0^2 increase depends on channel range and it increases more than several times when L increases from 100 to 700 m for constant C_n^2.

At the same time, as follows from Figure 5.5, where the same dependence is shown for different wavelengths, with the increase of the intensity of turbulence strength for constant channel range L and different wavelength λ from 0.1 to 0.7 μm, β_0^2 increases linearly. As also follows from Figure 5.5, with the increase of the wavelength from

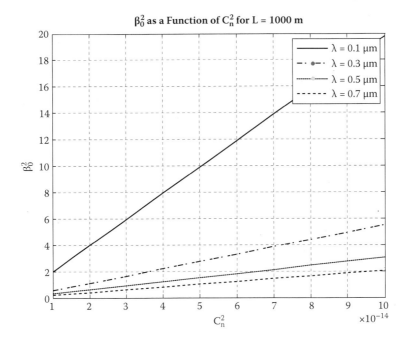

Figure 5.5 β_0^2 as a function of C_n^2 for different wavelengths; range $L = 1$ km.

0.1 to 0.7 μm for a constant link range of $L = 1$ km, the signal fluctuation index, β_0^2, decreases significantly, because with the increase of wavelength of optical waves, the efficiency of the scattering effect of rays becomes negligible.

It is interesting now to analyze BER as a function of SNR recorded at the optical receiver (or detector). Figure 5.6 shows scatterplots of the expected BER results versus the values of SNR, according to Equation 5.14, yielded from the experimental measurement data taken from References 9 and 36. The bottom curve in Figure 5.6 corresponds to the weakest level of atmospheric turbulence, usually observed at nighttime. The upper curve corresponds to daytime measurements, when the strongest levels of atmospheric turbulence were obtained. The nominal values of the refractive index structure parameter at nighttime and daytime were about $C_n^2 \approx 10^{-15}$ m$^{-2/3}$ and $C_n^2 \approx 10^{-13}$ m$^{-2/3}$, respectively. The dashed curves in Figure 5.6 correspond to intermediate levels of turbulence, plotted with steps of $\Delta C_n^2 \approx 3.5 * 10^{-15}$ m$^{-2/3}$, measured along 24 hours (regardless of the time of day).

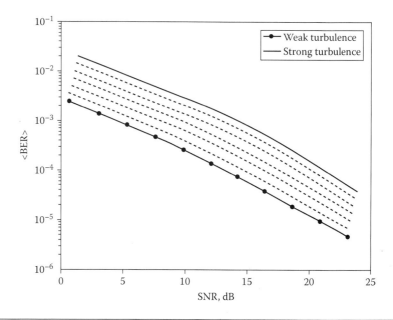

Figure 5.6 Plots of the expected BER results versus values of SNR, yielded from experimental measurement data for a 3.76 km propagation path.

Upon the same experimental values of the refractive index structure parameters, which describe atmospheric turbulence, additional evaluations of expected BER versus SNR for different propagation paths were made. Evaluation results evolve from applying Equation 5.14 for propagation paths of 5, 2, and 0.5 km, shown in Figure 5.7a, b, and c, respectively, by using experimental data obtained in References 9 and 36.

As seen from Figure 5.7 and others, an increase in propagation distance between the transmitter and the receiver systems yields an increase in expected BER results as expected from longer paths through. In addition, gaps between weak and strong turbulence curves become smaller.

A comparison of BER and SNR curves over extreme experimental turbulence conditions ($C_n^2 \approx 10^{-15} \text{ m}^{-2/3}$ and $C_n^2 \approx 10^{-13} \text{ m}^{-2/3}$ for daytime and nighttime, respectively) for different propagation lengths is shown in Figure 5.8. According to this plot, differences between the daytime curves for the same atmospheric turbulence value are small, in comparison with the nighttime curves gap.

As seen from Figure 5.8, strong turbulence atmospheric conditions, defined by $C_n^2 \approx 10^{-13} \text{m}^{-2/3}$, result in relatively great BER and,

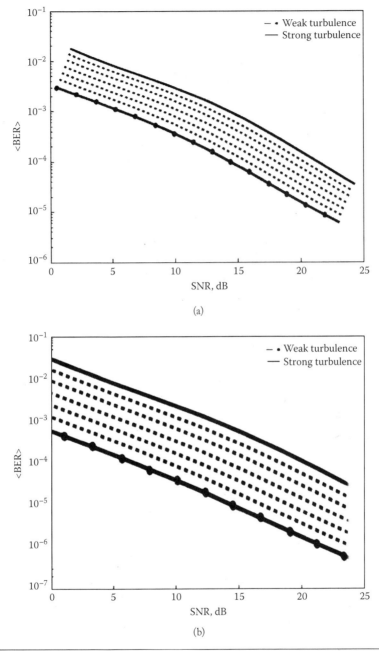

Figure 5.7   Evaluation of BER versus SNR values, corresponding to different atmospheric turbulence levels, for (a) 5 km propagation path, (b) 2 km propagation path, and (c) 0.5 km propagation path.

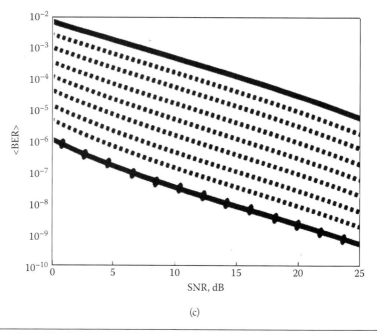

(c)

Figure 5.7   (*Continued*)

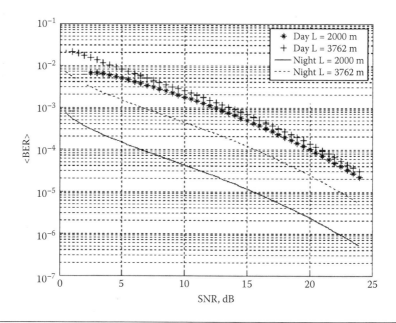

Figure 5.8   Comparison between 2 km and 3.76 km propagation lengths for extreme daytime and nighttime experimental turbulence conditions.

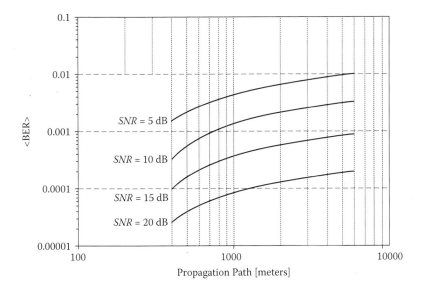

Figure 5.9   Representation of BER in strong atmospheric turbulence media as a function of propagation length, *L*, for different SNR values.

so, play a major role in fading effects, caused by turbulence [9, 19]. Therefore, there are subjects of particular interest for estimation of optical communication parameters, such as BER and SNR and prediction of maximal losses, caused by strong turbulence.

On the basis of experimental data taken from Reference 9, Figure 5.9 provides evaluation of BER in strong atmospheric turbulence media as a function of propagation length, *L*, for different SNR values. Communication link traces were taken from 0.4 to 6 km. For less than 0.4 km paths the error is negligible, and for over 6 km paths, optical communication systems (lasers) are not usually applied.

To explain the results obtained, we develop the following model for practical prediction of BER as a function of propagation path length and specific signal-to-noise values of wireless atmospheric optical communication links, for strong turbulence media:

$$\text{BER} = \varepsilon \ln L - \delta \qquad (5.34)$$

where *L* (in meters) is the propagation path length, and ε and δ are estimated signal-to-noise dependent factors, given in Figure 5.10.

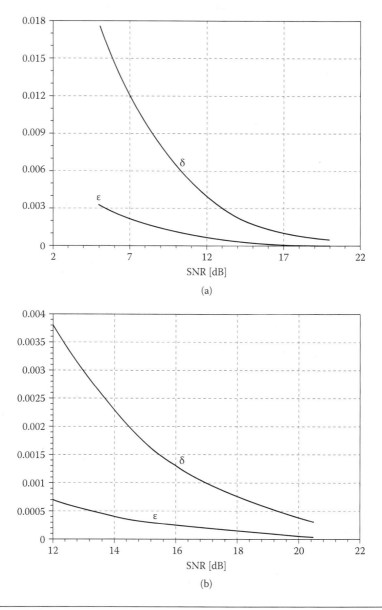

Figure 5.10   Diagram of ε and δ factors versus SNR values: (a) general view from 5 to 21 dB, and (b) close-up view from 12 to 21 dB.

Validation conditions for this model are as follows: (a) strong atmospheric turbulence media; (b) propagation path length from 0.4 to 6 km; (c) signal-to-noise values from 4 to 21 dB; and (d) BER from $2.5 \cdot 10^{-5}$ to 10^{-2}.

To present the capacity or spectral efficiency (for known bandwidth of the channel), we first need to determine relations between the scintillation index, σ_I^2, and β_0^2, which is the parameter that corresponds to the level of turbulence. The parameter $\beta_0^2 \approx 0.1$ for weak fluctuations and ≈ 4 for strong fluctuations. Using Equations 5.10–5.13, we have created a graph that demonstrates the relations between β_0^2 and σ_I^2 (see Figure 5.11).

As can be seen from Figure 5.11, with the growth of β_0 (i.e., stronger turbulence perturbations), the scintillation index that describes the signal fading grows. One can also observe the vast difference between the graphs for $d = 0$ (point receiver) and $d = 9$ (large collimating lens). The increase of d allows collection of all rays after scattering by turbulence along the wave path. For $d = 0$ the scintillation index grows rapidly as β_0 grows (stronger fluctuations),

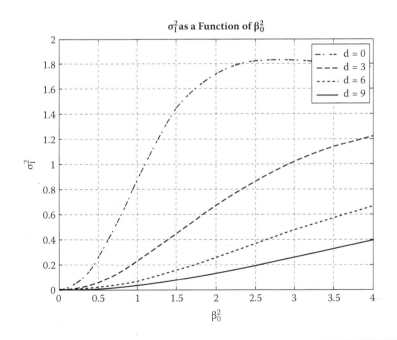

Figure 5.11 σ_I^2 as a function of β_0^2.

and for $d = 9$ the growth is much more moderate and is less sensitive to fluctuation. This can be explained by the more precise "focusing effect" of the initial beam passing through turbulent media by use of a detector with a collimated lens. Now, using Equation 5.33, we simulate the K parameter as a function of σ_I^2 in Figure 5.12.

From Figure 5.12a for $d = 0$ and Figure 5.12b for $d = 10$, we see that as the fluctuations grows, $K \to 0$; i.e., the channel tends to become a multipath Rayleigh channel. We can see from Figure 5.12a that for a small scintillation index (less than 0.1), the value of K is about 10 but drops rapidly as σ_I^2 increases. On the other hand, for $d = 10$ the initial value is much larger; i.e., there are better conditions for a communication channel. In this case, there is a great deterioration as σ_I^2 grows, but it does not tend to zero as fast as in the case of $d = 0$.

Using Equation 5.31, we can now evaluate the spectral efficiency as a function of K. The results are shown in Figure 5.13a and b for $d = 0$, where β_0 starts from 0 and from 0.1, respectively. As expected and observed for radio propagation [34], for an optical channel we find that as K grows the efficiency grows to unity as well.

It can be seen that for the case in which there are no fluctuations over the channel, as where β_0 starts from 0 (see Figure 5.13a), the capacity grows rapidly, and larger values of K are achieved, where we observe "saturation" of spectral efficiency for constant bandwidth β_w.

This can be explained as follows. With $K \geq 10$, the effects of turbulence are very small (i.e., the channel is close to the ideal case). An increase in a coherent component of the signals as compared to an incoherent one cannot change the effects of propagation conditions on the ideal channel with $K \geq 10$. When fluctuations are present all the time ($\beta_0 > 0$, Figure 5.13b), even very small ones, the efficiency does not grow as rapidly as observed earlier, and saturation is observed for smaller values of K. The same tendency is observed for $d = 10$, as presented in Figure 5.14.

The same tendency of saturation of spectral efficiency to unity is observed for smaller values of K due to a collimating effect of the lens at the receiver. Thus, if there is a receiver with a collimating lens ($d > 5$), even if K is small, i.e., the multiple scattering effects from turbulence are sufficiently large that, due to focusing effects, saturation is observed at the smaller K values.

K-Parameter as a Function of σ_I^2 for Point Receiver

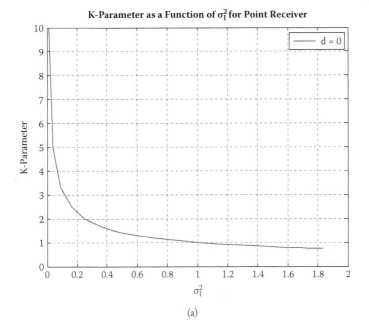

(a)

K-Parameter as a Function of σ_I^2 for Large Collimating Lens Receiver

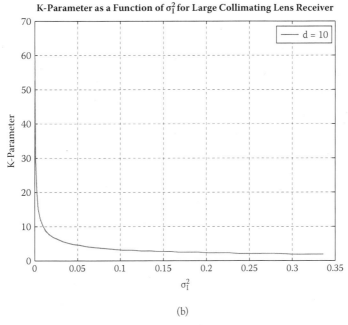

(b)

Figure 5.12 K as a function of σ_I^2 for (a) $d = 0$ and (b) $d = 10$. (From A. Tiker, N. Yarkoni, N. Blaunstein, A. Zilberman, and N. Kopeika, "Prediction of data stream parameters in atmospheric turbulent wireless communication links," *Appl. Opt.*, vol. 46, no. 2, pp. 190–199, 2007. With Permission.)

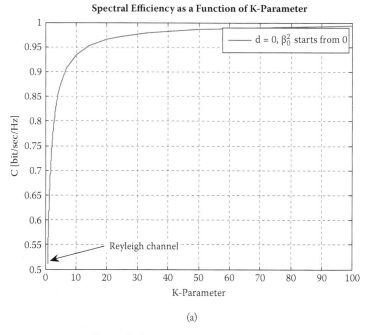

(a)

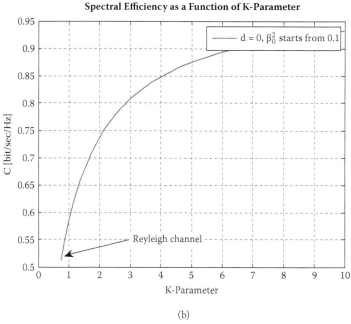

(b)

Figure 5.13 Spectral efficiency for $d = 0$ (point receiver): (a) β_0^2 starts from 0; (b) β_0^2 starts from 0.1. (From A. Tiker, N. Yarkoni, N. Blaunstein, A. Zilberman, and N. Kopeika, "Prediction of data stream parameters in atmospheric turbulent wireless communication links," *Appl. Opt.*, vol. 46, no. 2, pp. 190–199, 2007. With Permission.)

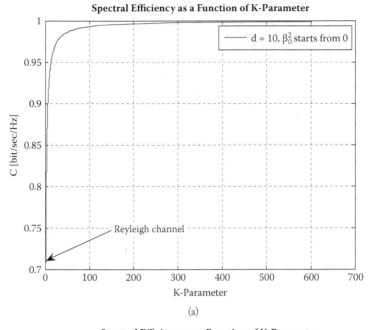

(a)

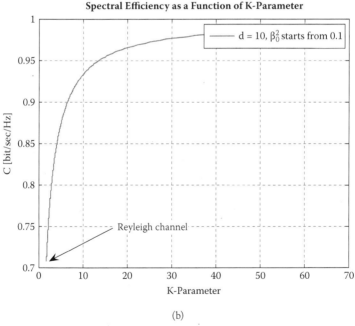

(b)

Figure 5.14   Spectral efficiency for $d = 10$ (large collimating lens): (a) β_0^2 starts from 0; (b) β_0^2 starts from 0.1. (From A. Tiker, N. Yarkoni, N. Blaunstein, A. Zilberman, and N. Kopeika, "Prediction of data stream parameters in atmospheric turbulent wireless communication links," *Appl. Opt.*, vol. 46, no. 2, pp. 190–199, 2007. With Permission.)

Now, we investigate the spectral efficiency of the atmospheric channel versus the Ricean parameter K for various SNRs. Results of computations are shown in Figure 5.15a and b for a point and collimated lens receiver, respectively.

It is clearly seen that, with an SNR increase from 1 to 16 dB, the spectral efficiency can be increased 3–5 times. The effect depends on the fading characteristics of the channel; i.e., on parameter K. It is clear that for $K > 5$–10, the LOS component prevails and a further increase of K parameter does not change the spectral efficiency of the channel. This tendency is observed both for the point and for the collimated lens receiver.

Finally, we analyze the effects of fading (e.g., the changes of the K parameter) on BER conditions within the turbulent wireless communication link. The results are shown in Figure 5.16a, for both a point receiver and a receiver with collimating lens (i.e., for d changing up to 9). It is clearly seen that with decrease of fading inside the channel (i.e., with increase of K parameter), the BER parameter is decreased sharply. Moreover, as follows from the illustrations, the rate of decrease of BER increases with an increase of the collimating parameter, d, of the receiver.

To compare channels, the BER computed for the horizontal path of 1 km at heights of 1–2 km is shown in Figure 5.16b, taking into account data for C_n^2 based on the computed mean value of $\langle C_n^2 \rangle = 4 \cdot 10^{-16}$ m$^{-2/3}$ [36].

As seen from the results of the computations presented in Figure 5.16b for the height range of 1–2 km, the same tendency of BER sharp decrease with increase of K parameter, obtained at lower altitudes of 100–200 m (see Figure 5.16a), is observed here too.

However, at higher altitudes due to weak turbulence ($\langle C_n^2 \rangle = 4 \cdot 10^{-16}$ m$^{-2/3}$ at 1–2 km compared to $\langle C_n^2 \rangle = 5 \cdot 10^{-14}$ m$^{-2/3}$ at 100–200 m), for the horizontal atmospheric channels where the LOS component exceeds the multipath (NLOS) component (i.e., for $K > 1$), the BER characteristic becomes negligible and can be ignored as well as other fading characteristics in the design of optical atmospheric links between stratospheric platforms.

According to the results, presented in Figure 5.16a and b, the optimal algorithm for the minimization of BER was performed for different situations occurring over optical atmospheric communication

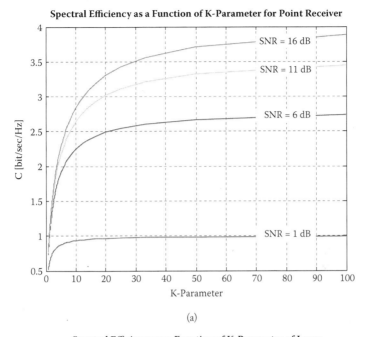

(a)

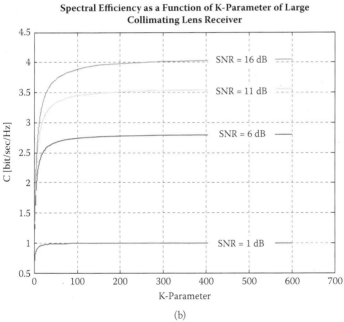

(b)

Figure 5.15   Spectral efficiency as a function of *K* parameter (SNR varying) for (a) point receiver and (b) collimated lens receiver (*d* = 10). Range *L* = 1 km. (From A. Tiker, N. Yarkoni, N. Blaunstein, A. Zilberman, and N. Kopeika, "Prediction of data stream parameters in atmospheric turbulent wireless communication links," *Appl. Opt.*, vol. 46, no. 2, pp. 190–199, 2007. With Permission.)

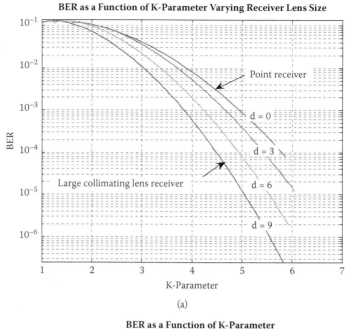

(a)

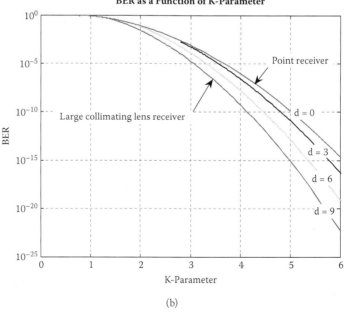

(b)

Figure 5.16   BER as a function of K parameter for varying receiver lens sizes for heights of (a) 100–200 m and (b) 1–2 km. Range $L = 1$ km. (From A. Tiker, N. Yarkoni, N. Blaunstein, A. Zilberman, and N. Kopeika, "Prediction of data stream parameters in atmospheric turbulent wireless communication links," *Appl. Opt.*, vol. 46, no. 2, pp. 190–199, 2007. With Permission.)

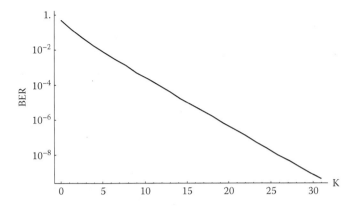

Figure 5.17   BER as a function of K.

links. Thus, taking some measured data, presented in References 27 and 30, we can show here some examples. In our computations we used the following parameters of the channel and measured data: $\sigma = 2$ dB and $SNR_{add} = 1$ dB. The results of the computations are shown in Figure 5.17 for BER as a function of the fading parameter K obtained from the experimental data.

As seen from Figure 5.17, with an increase of the K parameter (that is, when the LOS component becomes predominant with respect to the NLOS multipath components), it is found that BER decreases essentially from 10^{-2} for $K \approx 5$ to 10^{-6} for $K \approx 20$ (i.e., for the atmospheric link at altitudes of 100–500 m filled by turbulent structures [27, 30]). At the same time, as expected in Reference 35, the spectral efficiency also is found to be increased with the K parameter. Hence, with the increase of the spectral efficiency of the data stream (from 0.8 to 1.0), a simultaneous sharp decrease of BER was also found.

5.4 Summary

In this chapter, we described and analyzed a unified approach for calculation of signal data parameters in the atmospheric optical communication channel with fast fading caused by atmospheric turbulence. The effects of turbulence are described via the well-known Kolmogorov model of turbulence relaxation in terms of random signal scintillations described by the gamma-gamma distribution combined with the generalized Ricean K parameter. These effects were obtained based on measurements of the values of the refractive index structure parameter, C_n^2, presented in References 9 and 36. The relation between the Ricean parameter, K, and the signal scintillation parameter, σ_I^2, was considered to develop a unified description of the corresponding PDF of signal fading in an atmospheric wireless communication link. We derived and estimated the corresponding PDF and the K parameter, and then data parameters such as the SNR, the BER, and the capacity of the optical atmospheric channel (C). Such an approach allows us to permit reliable prediction of effects of fading caused by different levels of turbulence. Comparison with experimental data, observed at different heights, from the lower atmospheric level of 100–200 m to the high atmospheric level above 1–2 km (described in References 9 and 36) is presented. It was shown that close to the ground surface, at heights of 100–200 m, the effects of fading caused by turbulence are much greater compared with those observed at heights of 1–2 km. This is because of weaker turbulences occurring at higher altitudes and stronger turbulences occurring at lower altitudes. As was shown from Figures 5.6–5.8, strong atmospheric turbulence leads to much greater BER and SNR values, compared with those observed for weak turbulence. Therefore, the data stream characteristics, such as the channel capacity and the efficiency, become larger in optical atmospheric links designed for higher altitudes, and the BER characteristics become negligible and can be ignored.

References

1. T. H. Nielsen, "IPv6 for wireless networks," *J. Wireless Personal Commun.*, vol. 17, pp. 237–247, 2001.
2. S. Ohmori, Y. Yamao, and N. Nakajima, "The future generation of mobile communications based on broadband access methods," *Wireless Personal Communication*, vol. 17, pp. 175–190, 2001.
3. G. M. Djuknic, J. Freidenfelds, and Y. Okunev, "Establishing wireless communication services via high-altitude aeronautical platforms: A concept whose time has come?" *IEEE Commun. Magazine*, pp. 128–135, 1997.
4. Y. Hase, R. Miura, and S. Ohmori, "A novel broadband all-wireless access network using stratospheric radio platform," VTC'98 (48th Vehicular Technology Conf.), Ottawa, Canada, May 1998.

5. P. Greiling and N. Ho, "Commercial satellite applications for heterojunction microelectronics technology," *IEEE Trans. MTT*, vol. 46, no. 6, pp. 734–738, 1998.

6. S. R. Saunders, *Antennas and Propagation for Wireless Communication Systems*, Wiley & Sons, New York, 1999.

7. L. C. Andrews and R. L. Phillips, *Laser Beam Propagation through Random Media*, SPIE Optical Engineering Press, Bellingham, WA, 1998.

8. F. G. Stremler, *Introduction to Communication Systems*, Addison-Wesley, Reading, MA, 1982.

9. N. S. Kopeika, *A System Engineering Approach to Imaging*, SPIE Optical Engineering Press, Bellingham, WA, 1998.

10. V. A. Banakh and V. L. Mironov, *LIDAR in a Turbulence Atmosphere*, Artech House, Dedham, MA, 1987.

11. W. Zhang, J. K. Tervonen, and E. T. Salonen, "Backward and forward scattering by the melting layer composed of spheroidal hydrometeors at 5–100 GHz," *IEEE Trans. Antennas Propagat.*, vol. 44, pp. 1208–1219, 1996.

12. A. Macke and M. Mishchenko, "Applicability of regular particle shapes in light scattering calculations for atmospheric ice particles," *Appl. Opt.*, vol. 35, pp. 4291–4296, 1996.

13. E. A. Hovenac, "Calculation of far-field scattering from nonspherical particles using a geometrical optics approach," *Appl. Opt.*, vol. 30, pp. 4739–4746, 1991.

14. J. D. Spinhirme and T. Nakajima, "Glory of clouds in the near infrared," *Appl. Opt.*, vol. 33, pp. 4652–4662, 1994.

15. L. D. Duncan, J. D. Lindberg, and R. B. Loveland, "An empirical model of the vertical structure of German fogs," ASL-TR-0071, U.S. Army Atmospheric Sciences Laboratory, White Sands Missile Range, NM, 1980.

16. L. C. Andrews, R. L. Phillips, and C. Y. Hopen, *Laser Beam Scintillation with Applications*, SPIE Optical Engineering Press, Bellingham, WA, 2001.

17. A. Ishimaru, *Wave Propagation and Scattering in Random Media*, Academic Press, New York, 1978.

18. A. S. Monin and A. M. Obukhov, "Basic law of turbulent mixing near the ground," *Trans. Akad. Nauk.*, vol. 24, no. 151, pp. 1963–1987, 1954.

19. V. I. Tatarskii, *Wave Propagation in a Turbulent Medium*, McGraw-Hill, New York, 1961.

20. D. Dion and P. Schwering, "On the analysis of atmospheric effects on electro-optical sensors in the marine surface layer," *Proc. NATO-IRIS Conf.*, vol. 41, no. 3, pp. 305–322, June 1996.

21. A. Berk, L. S. Bernstein, and D. C. Robertson, "MODTRAN: A moderate resolution model for LOWTRAN 7," Tech. rep. GL-TR-89-0122, Air Force Geophysics Laboratory, Hanscom Air Force Base, Bedford, MA, 1989.

22. R. R. Beland, "Propagation through atmospheric optical turbulence," in Atmospheric propagation of radiation, F. G. Smith, Ed., in *The Infrared and Electro-Optical Systems* J. S. Accetta and D. L. Shumaker, Executive Eds., vol. 2, SPIE Optical Engineering Press, Bellingham, WA, pp. 157–232, 1993.

23. V. Thiermann and A. Kohnle, "A simple model for the structure constant of temperature fluctuations in the lower atmosphere," *J. Phys.*, vol. 21, S37–S40, 1988.

24. D. Sadot and N. S. Kopeika, "Forecasting optical turbulence strength on basis of macroscale meteorology and aerosols: Models and validation," *Optical Eng.*, vol. 31, pp. 200–212, 1992.

25. D. L. Hutt, "Modeling and measurements of atmospheric optical turbulence over land," *Optical Eng.*, vol. 38, no. 8, pp. 1288–1295, 1999.

26. J. S. Accetta and D. L. Shumaker, Eds., *The infrared and Electro-Optical Systems*, SPIE Optical Engineering Press, Bellingham, WA, pp. 157–232, 1993.

27. N. S. Kopeika, I. Kogan, R. Israeli, and I. Dinstein, "Prediction of image quality through atmosphere as a function of weather forecast," in *Propagation Engineering*, N. S. Kopeika and W. B. Miller, Eds., Proc. SPIE, vol. 1115, pp. 266–277, 1989.

28. A. N. Kolmogorov, "The local structure of turbulence incompressible viscous fluid for very large Reynolds numbers," *Reports Acad. Sci. USSR*, vol. 30, pp. 301–305, 1941.

29. C. Y. Young, Y. V. Gilchrest, and B. R. Macon, "Turbulence induced beam spreading of higher order mode optical waves," *Optical Eng.*, vol. 41, no. 5, pp. 1097–1103, 2002.

30. N. S. Kopeika, I. Kogan, R. Israeli, and I. Dinstein, "Prediction of image propagation quality through the atmosphere: The dependence of atmospheric modulation transfer function on weather," *Optical Eng.*, vol. 29, no. 12, pp. 1427–1438, 1990.

31. L. C. Andrews, *Special Functions of Mathematics for Engineers*, 2nd ed., SPIE Optical Engineering Press, Bellingham, WA; Oxford University Press, Oxford, 1998 (formerly published as 2nd ed. by McGraw-Hill, New York, 1992).

32. A. Papoulis, *Probability Random Variables and Stochastic Process*, McGraw-Hill, New York, 1991.

33. J. G. Proakis, *Digital Communication*, 4th ed., McGraw-Hill, New York, 2001.

34. N. Blaunstein and C. Christodoulou, *Radio Propagation and Adaptive Antennas for Wireless Communication Links: Terrestrial, Atmospheric, and Ionospheric*, Wiley InterScience, Hoboken, NJ, 2007.

35. N. Yarkoni and N. Blaunstein, "Capacity and spectral efficiency of MIMO wireless systems in multipath urban environments with fading," *Proc. of European Conf. on Antennas and Propagation*, Nice, France, 6–10 November 2006, pp. 316–321.

36. A. Tiker, N. Yarkoni, N. Blaunstein, A. Zilberman, and N. Kopeika, "Prediction of data stream parameters in atmospheric turbulent wireless communication links," *Appl. Opt.*, vol. 46, no. 2, pp. 190–199, 2007.

37. E. Biglieri, J. Proakis, and S. Shamai, "Fading channels: Information-theoretic and communication aspects," *IEEE Trans. Information Theory*, vol. 44, no. 6, pp. 2619–2692, 1998.

38. M.-S. Alouini, M. K. Simon, and A. J. Goldsmith, "Average BER performance of single and multi carrier DS-CSMA systems over generalized fading channels," *Wiley J. Wireless Systems Mobile Comput.*, vol. 1, no. 1, pp. 93–110, 2001.

39. A. J. Goldsmith, L. J. Greenstein, and G. L. Foschini, "Error statistics of real-time power measurements in cellular channels with multipath and shadowing," *IEEE Trans. Vehicular Technol.*, vol. 43, no. 3, pp. 439–446, 1994.

40. A. J. Goldsmith, "The capacity of downlink fading channels with variable rate and power," *IEEE Trans. Vehicular Technol.*, vol. 46, no. 3, pp. 569–580, 1997.

41. A. J. Goldsmith and P. P. Varaiya, "Capacity of fading channels with channel side information," *IEEE Trans. Information Theory*, vol. 43, no. 6, pp. 1986–1992, 1997.

42. I. E. Telatar and D. N. C. Tse, "Capacity and mutual information of wide-band multipath fading channels," *IEEE Trans. Information Theory*, vol. 46, no. 4, pp. 1384–1400, 2000.

43. G. L. Stuber, *Principles of Mobile Communication*, Kluwer Academic Publishers, 1996.

44. N. Blaunstein, N. Yarkoni, and D. Katz, "Spatial and temporal distribution of the VHF/UHF radio waves in built-up land communication links," *IEEE Trans. Antennas Propagation*, vol. 54, no. 8, pp. 2345–2356, 2006.

45. J. H. Winters, "On the capacity of radio communication systems with diversity in a Rayleigh fading environment," *IEEE J. Selected Areas Commun.*, vol. 5, pp. 871–878, 1987.

Abbreviations

1D	one-dimensional
2D	two-dimensional
3D	three-dimensional
ABL	atmospheric boundary layer
ACF	autocorrelation function
ADSL	asynchronous digital subscriber line
AFGL	Air Force Geophysical Laboratory
AOA	angle of arrival
ASD	aerosol size distribution
ASK	amplitude shift keying
AWGN	additive white Gaussian noise
BEP	bit error probability
BER	bit error rate
CATV	cable television
CCD	charge-coupled device
CDF	cumulative distribution function
CLT	central limit theorem
CW	continuous wave
dB	decibels
DF	distance-frequency
DSNR	digital signal-to-noise ratio
EM	electromagnetic

erfc (*)	error function
FFT	fast Fourier transform
FM	frequency modulated
FOV	field of view
FSK	frequency shift keying
FSO	free-space optics
GOA	geometric optical approximation
GoS	grade of service
GSM	generalized seeing monitor
HF	high frequency
H-V (HV)	Hufnagel-Valley
ICCD	intensified charge-coupled device
i.i.d.	independent, identically distributed
IR	infrared
K	Ricean parameter (factor)
LED	light-emitting diode
LFM	linear frequency modulation
LGS	laser guide star
LIDAR	light detection and ranging
LOS	line-of-sight
LT	local time
MAP	maximum a posteriori probability
MLE	maximum likelihood estimate/estimator
MODTRAN	moderate resolution model
MOS	macroscale optical simple
NLOS	non-line-of-sight
N-O-K	non-Obukhov-Kolmogorov
O-K	Obukhov-Kolmogorov
OOK	on-off keying
OWC	optical wireless communication
PAM	pulse amplitude modulation
PC	personal computer
PDF	probability density function
PFC	phase frequency characteristic
PIN	positive intrinsic negative
PLC	power line communication
PPM	pulse position modulation
PSD	power spectral density

PSK	phase shift keying
RH	relative humidity
RMS	root mean square
S4	index of scintillation
SCIDAR	scintillation detection and ranging
SLC	submarine laser communication
SNR	signal-to-noise ratio
TD	time delay
UT	universal time
UV	ultraviolet

Index